Fifth Edition

Our Digital World

Introduction to Computing

Karen Lankisch • Nancy Muir
Denise Seguin • Anita Verno

PARADIGM
EDUCATION SOLUTIONS

St. Paul

Vice President, Content and Digital Solutions: Christine Hurney
Director of Content Development, Computer Technology: Cheryl Drivdahl
Director of Production: Timothy W. Larson
Developmental Editor: Katie Werdick
Associate Production Editor/Project Manager: Melora Pappas
Cover and Text Design: Jaana Bykonich
Senior Design and Production Specialist: Julie Johnston
Instructional Support Writer: Jeff Johnson
Copy Editor: Sarah Kearin
Indexer: Terry Casey
Vice President, Digital Solutions: Chuck Bratton
Digital Projects Manager: Tom Modl
Digital Solutions Manager: Gerry Yumul
Vice President, Sales and Marketing: Scott Burns
Senior Director of Marketing: Lara Weber McLellan

Care has been taken to verify the accuracy of information presented in this book. However, the authors, editors, and publisher cannot accept responsibility for web, email, newsgroup, or chat room subject matter or content, or for consequences from the application of the information in this book, and make no warranty, expressed or implied, with respect to its content.

Trademarks: Microsoft is a trademark or registered trademark of Microsoft Corporation in the United States and/or other countries. Some of the product names and company names included in this book have been used for identification purposes only and may be trademarks or registered trade names of their respective manufacturers and sellers. The authors, editors, and publisher disclaim any affiliation, association, or connection with, or sponsorship or endorsement by, such owners.

Paradigm Publishing, Inc., is independent from Microsoft Corporation and not affiliated with Microsoft in any manner.

Cover Photo Credit: © zffoto/Shutterstock.com
Interior Photo Credits: Following the index.

We have made every effort to trace the ownership of all copyrighted material and to secure permission from copyright holders. In the event of any question arising as to the use of any material, we will be pleased to make the necessary corrections in future printings.

ISBN 978-0-76388-679-0 (print)
ISBN 978-0-76388-678-3 (digital)

© 2020 by Paradigm Publishing, Inc.
875 Montreal Way
St. Paul, MN 55102
Email: CustomerService@ParadigmEducation.com
Website: ParadigmEducation.com

Printed in the United States of America

28 27 26 25 24 23 22 21 20 19   1 2 3 4 5 6 7 8 9 10

Brief Contents

Contents

Preface

Getting the Most Out of *Our Digital World*

You've just paid good money for another textbook. It seems like every other textbook. You expect it to be something you have to get through, and something you'll be glad to leave behind when you finish the course.

The truth is, we actually hope you are somewhat surprised by this book.

As in the previous editions of *Our Digital World*, we've done a lot of work to make the writing in this book easy to read, to find ways to get you excited about how technology is evolving, and to make computing relevant to your work and personal life. We've included information about recent technologies such as cloud computing and mobile applications. We've also cut back on the amount you have to read by providing part of the content in multimedia formats that we think you'll find engaging. The result is more than a book—it's a combination of text and technology that together, provides a new learning experience.

The multimedia content of *Our Digital World,* Fifth Edition, is available as part of Cirrus, a web-based training and assessment system.

Moving Your Learning Online

One of the fundamental ways that this book provides a different learning experience is by connecting this course with the way you experience computing in the twenty-first century. We have integrated the use of online technology into this textbook through a web-centric educational experience. The activities involve accessing the Cirrus online course to watch videos, play with interactive hands-on tools, and connect with other students and your instructor by learning how to use collaborative features such as blogs and wikis. To see how this works, take a look at the Chapter Tour beginning on page x.

Our hope is that you'll not only find our online features informative and interesting, but that you'll also become a more competent participant in our digital world by gaining practice with online technologies. After you finish the course, that practice will continue to help you enjoy computers on a personal and professional level.

Pay special attention to the Take the Next Step activities that are marked at the end of chapter sections. These activities, available in the Cirrus online course, provide required learning in an alternative and visual way. A variety of other Cirrus online activities—including videos, surveys, and blogs—are provided to enhance your learning in ways that suit your class environment and interests.

Who Are You?

We know that our students are not all the same, so we've tried to address a variety of interests and backgrounds. You are 18 or 28 or 60 years old. You are comfortable with technology or you may be technophobic. You could be juggling a job and family with your education to take the next step in your career, or you could be just starting out on your initial career path. Maybe you recently retired and are looking for a new work experience.

This course could be a prerequisite for your degree in a subject unrelated to computing, or it could be a stepping stone to a career that is focused on computing. Maybe it's simply going to help you keep up with the digital curve so you can explore ways that computers can connect you with friends and family.

Whatever your goals, we believe this courseware will help you, and we wish you success in exploring the amazing possibilities of our digital world.

Taking a Chapter Tour

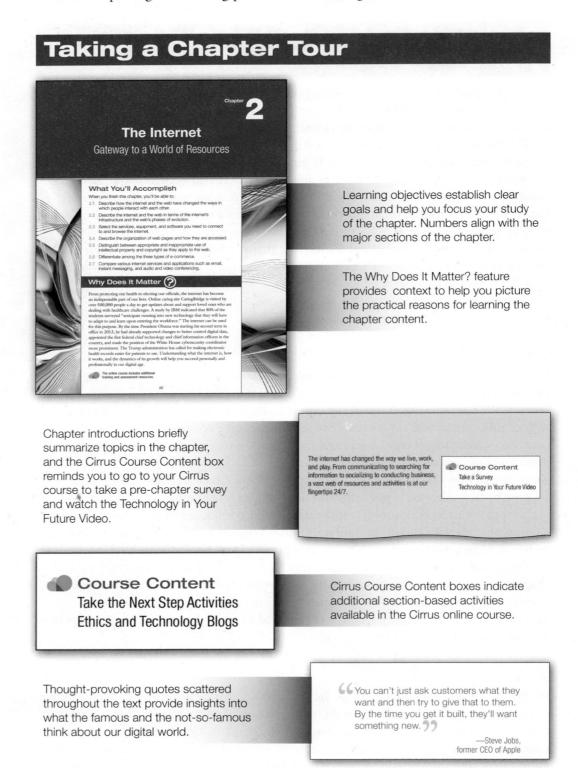

Learning objectives establish clear goals and help you focus your study of the chapter. Numbers align with the major sections of the chapter.

The Why Does It Matter? feature provides context to help you picture the practical reasons for learning the chapter content.

Chapter introductions briefly summarize topics in the chapter, and the Cirrus Course Content box reminds you to go to your Cirrus course to take a pre-chapter survey and watch the Technology in Your Future Video.

Cirrus Course Content boxes indicate additional section-based activities available in the Cirrus online course.

Thought-provoking quotes scattered throughout the text provide insights into what the famous and the not-so-famous think about our digital world.

> "You can't just ask customers what they want and then try to give that to them. By the time you get it built, they'll want something new."
>
> —Steve Jobs, former CEO of Apple

TABLE 2.1 Three Phases of the Web

Phase	Also Known As	User Capabilities
Web 1.0	Static web	Read content but can't interact with anything
Web 2.0	Interactive web	Read content, interact with content, write content, collaborate
Web 3.0	Semantic web	Read content, interact with content, write content, collaborate, access connected content, use devices that can talk to one another to gather information and present useful information

Tables organize data and summarize concepts.

Computers in Your Career

The internet and web have made available a wide variety of careers that didn't exist 10 years ago, and not all are high-tech in nature. For example, consider jobs such as web content writer, internet law expert, and online trainer. If you are more technically inclined, a variety of developer jobs allow you to work with tasks from programming an e-commerce shopping cart to creating environments in virtual reality worlds.

Computers in Your Career features give ideas for how your computer studies could help you succeed in a career you may not have considered.

FIGURE 2.12 The B2C Online Shopping Process

Illustrations help you visualize processes and concepts explained in the text. Even if you're not a visual learner, you may find that a picture can save you reading a thousand words.

Step 1 Customer visits an electronic storefront and views the online catalog.

Step 2 Items are selected and placed into the shopping cart.

NAME: ********
ADDRESS: *********
CREDIT CARD: ********

Step 3 Customer accesses checkout and enters personal and financial data.

Step 5 Customer receives confirmation notice of the purchase.

bank's server

Step 4 E-tailer verifies shopper's financial data at a banking website.

e-tailer's secure server

Playing It Safe

Most people don't realize how much information about them exists online. Maybe you posted a résumé or made some blog entries on a publicly viewable page. Perhaps schools, employers, friends, or the government placed your information online. Do a web search of your name. If you find information about you that you'd rather not have online, ask the host site to remove it.

Playing It Safe margin boxes provide practical tips for using the web safely and responsibly.

The Review and Assessment section begins with Summing Up, a recap of the chapter concepts. The Cirrus Course Content box reminds you to go to your Cirrus course to access additional training and assessment resources.

Review and Assessment

Course Content
The online course includes additional training and assessment resources.

Summing Up

2.1 The World Goes Online

Today more people are using the internet for a variety of activities over several kinds of devices, such as smartphones and tablets. You can **sync** devices to share data for calendars, contacts, and more.

There are many examples of cutting-edge uses of the internet in fields such as science, medicine, and even space travel. People are influenced in their purchasing decisions by the internet. **QR codes** are turning up everywhere for you to scan from your smartphone to get information about businesses and historical locations.

In the future, we will see even more uses of technology, more social collaboration, and advances in everything from retail to healthcare to communication from space, all because of the internet.

Terms to Know

2.1 The World Goes Online
sync, 26
smart speaker, 27
quick response (QR) code, 28

2.2 What Are the Internet and World Wide Web?
internet, 29
web, 29
web page, 29
website, 29
Web 2.0, 31
Web 3.0, 32
Semantic Web, 32
Internet of Things (IoT), 32

2.3 Joining the Digital World
internet service provider (ISP), 34
tethering, 35
multihomed device, 35
browser, 36
download, 36
upload, 36

2.4 Navigating and Searching the Web
Internet Protocol (IP) address, 37
uniform resource locator (URL), 37
web address, 37
hypertext, 38
home page, 38
search engine, 39

Page numbers in the Terms to Know section indicate where the term is bolded and defined within the chapter. To review the definitions of these terms, turn to the complete Glossary at the end of the book or in the Cirrus online course.

Completing the chapter Projects allows you to put your new knowledge to work, either on your own or as a member of a team. These activities let you demonstrate the kind of thinking and problem-solving needed in today's workplace—plus you can use popular social media, such as wikis, in the process. The projects can be found in your book or in the Cirrus course, where you can upload and submit files to your instructor.

Projects

Check with your instructor for the preferred method to submit completed work.

Smartphones—Staying Well While Staying Connected

Project 2A TEAM

Play the interview with Dr. Eric Topol at https://ODW5.ParadigmEducation.com/iDoctor. Consider the ways he suggests that smartphones can be used to replace expensive medical treatments, and then respond to the following:
1. List three uses for smartphones that are demonstrated by Dr. Topol in the video.
2. What does Dr. Topol say about the future possibility of using a smartphone app to replace a visit to the doctor?
3. Do you believe your smartphone can be used to replace your physician? Defend your answer.

Individually or in teams, prepare for a debate. Can smartphones replace physicians? Your instructor will assign you to the pro or con position.

Project 2B

Texting while driving is a safety hazard. Countries around the world have acted to help keep roads safe by enacting laws that prohibit texting while driving. According to the FCC (US) article at https://ODW5.ParadigmEducation.com/TextDriving, distracted driving kills more than 8 people and injures 1,161 daily. Texting and making

Class Conversations

Topic 2A   Is web research enough?

You are taking a sociology course and have been asked to write an essay on various kinds of bullying in high schools. Assume you conduct all of your research online, making sure that you use credible sources. Another classmate conducts all of her research at the library using traditional reference materials such as sociology journals and books.

Discuss the advantages of your internet-based research approach and the advantages of your classmate's library-based work. Should the two papers be considered equal when graded? Do you think others will view the journal- and book-researched paper as being of stronger academic quality? Why or why not? Is your web-researched paper likely to have a different perspective? Why or why not?

Optional: Create a blog entry that states your response to the previous questions and provides your rationale.

Topic 2B   How is YouTube changing society?

The popularity of YouTube has made it possible for anyone to become a media publisher and can make an average person suddenly famous. It seems that every week another video is posted that becomes a popular topic among friends and family members. Some of these videos are taken of other people with or without their knowledge.

Consider the following scenario. You are enjoying an afternoon with friends at a football game. You jump up to cheer and accidentally trip, causing the person next to you to dump a carton of popcorn all over you. You are not injured, but are somewhat embarrassed by the popcorn on your clothes and in your hair. Your friend, who thought you looked fu[...] filmed the accident using a cell

In the final Review and Assessment activities, you get a chance to talk about some key ideas and to share your thoughts on the effects of current and future technology applications. Take your discussion online (if your instructor sets this section up as a blog) or extend your conversations in class with the challenging questions presented in the Class Conversations section.

The Cirrus Solution

Our Digital World, Fifth Edition, is more than just a textbook. The textbook features and the digital content, available as part of your Cirrus course, are vital to your learning experience.

Powered by Paradigm, Cirrus is the next-generation learning solution for developing skills in Microsoft Office and computer concepts. Cirrus seamlessly delivers complete course content in a cloud-based learning environment that puts you on the fast-track to success. You can access your content from any device anywhere, through a live internet connection. Cirrus is platform independent, ensuring that all students get the same learning experience whether they are using a PC, Mac, or Chromebook computer. Cirrus provides access to all the *Our Digital World* content, delivered in a series of scheduled assignments that report to a grade book to track student progress and achievement.

Chapter Introduction

The chapter introduction activities establish the learning objectives for the chapter and provide a starting point for class discussion.

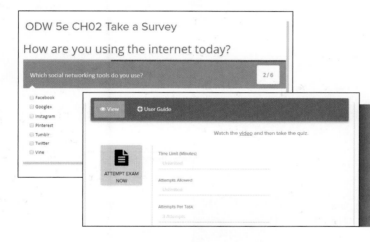

In Cirrus, Take a Survey activities invite you to answer a few questions on a topic related to the chapter. The Technology in Your Future Videos explore cutting edge technologies and computing concepts in videos followed by short comprehension quizzes.

Dynamic Training

The Cirrus online version of the *Our Digital World* course includes interactive assignments to guide student learning.

Watch and Learn Lessons offer opportunities to view the content for each chapter in a video presentation, read the content for each section, and complete a short multiple-choice quiz for each section.

Take the Next Step Activities expand on the content taught in the section and include a brief quiz.

Watch the video and then take the quiz.

5 Choices for Internet Connection

ATTEMPT EXAM NOW

Time Limit (Minutes)
Unlimited

Attempts Allowed

Question **1**

Not yet answered

Points out of 1.00

Flag question

Some people think the web should be regulated by a central body. Other people think the freedom to say and do what you want on the web should be protected. What do you think? Should the web be under anybody's control?

Paragraph

The Ethics and Technology Blog activities prompt you to blog about some of the ethical questions related to living and working with technology.

Chapter Review and Assessment

Chapter review and assessment activities in the Cirrus environment reinforce and assess student learning.

An interactive version of the Summing Up chapter summary provides pop-up definitions of the bolded chapter key terms.

The World Goes Online

Today more people are using the internet for a variety of activities over several kinds of devices, such as smartphones and tablets. You can sync devices to share data for calendars, contacts, and more.

There are many examples of cutting-edge uses of the internet in fields such as science, medicine, and even space travel. People are influ... QR codes are turning up everywhere for you to scan ... esses and historical locations.

In the future, we will see eve... and advances in everything from retail to hea... the internet.

What Are the Internet and

The internet is the physical ... at allows us to share resources and communicate ... are such as servers, routers, switches, transmission ... and transmit vast amounts of data.

The web is a body of content that is available as web pages. The pages are stored on servers around the world. A web page may contain text, images, interactive animations, games, music, and more. Several web

sync

The process of updating data on one device based on changes made to the data on another device. Short for synchronize.

OK

ODW 5e CH02 Flash Cards

File Transfer Protocol (FTP)

Turn

Terms to Know flash cards can help you review content before the chapter exam.

ODW 5e CH02 Concepts Check 2.4 Arrange It

Arrange the steps followed by data moving across the Internet.

Drag to arrange the images in the correct sequence

Passes through national ISP(s)	Information requested	Passes through routers	Domain name becomes IP address	Arrives at destination server	Request sent

ODW 5e CH02 Concepts Check 2.3 Label It

Label the components of a URL.

ODW 5e CH02 Concepts Check 2.3

`https://www.ParadigmEducation.com`

World Wide Web page	domain name	top-level domain name
	hypertext transfer protocol secure	

Cirrus activities review chapter content in fresh and engaging ways. Arrange It activities (above) prompt you to organize a series of images such as a sequence of events. Label It activities (left) ask you to identify different parts of an image such as a motherboard or a URL. Additional review activities include a multiple-choice quiz and a matching quiz.

Concept Exams with multiple-choice questions assess student understanding of the chapter content.

Question 1

Not yet answered

Points out of 1.00

Flag question

A 2017 study estimated that more than _____ billion people were online worldwide, up from almost 3.3 million people in 2015.

Select one:

a. 1

b. 6.5

c. 4.1

d. 15

Ebook

The entire book's contents is also available as an eBook in Cirrus from any device (desktop, laptop, tablet, or smartphone) anywhere. The eBook features dynamic navigation tools including a linked table of contents and the ability to jump to specific pages, search for terms, bookmark, highlight, and take notes.

Instructor eResources

Accessed through Cirrus and visible only to instructors, the Instructor eResources for *Our Digital World* include the following support:

- Planning resources, such as lesson plans, teaching tips, and sample course syllabi
- Delivery resources, such as PowerPoint® presentations with lecture notes
- Assessment resources, including answer keys and rubrics for evaluating chapter work and a concept item bank for each chapter that can be used to create custom exams

About the Author Team

Karen Lankisch

Professor, Consultant
University of Cincinnati—Clermont College
Cincinnati, Ohio

Karen Lankisch is a professor at the University of Cincinnati—Clermont College. Lankisch is the program director of the Health Information Systems Technology program and teaches a variety of courses in the Health and Business Information Technology programs. Lankisch is a Quality Matters Master Peer Reviewer and has completed reviews both nationally and internationally. She has a PhD in Education with a concentration in technology and adult learning and is certified through AHIMA as a Registered Health Information Administrator (RHIA) and Certified Health Data Analyst (CHDA). Along with co-authoring *Our Digital World*, Lankisch has co-authored *Exploring Electronic Health Records* for Paradigm Education Solutions. In 2013, she received the University of Cincinnati Faculty Award for Innovative Use of Technology in the Classroom. In 2016, she was selected for membership to the University of Cincinnati Academy of Fellows for Teaching and Learning. Dr. Lankisch was awarded the 2018 University of Cincinnati Clermont Mentoring Award.

Nancy Muir
Writer, Author, Instructor
Seattle, Washington

Nancy Muir is the owner of The Publishing Studio, a technology publishing company. Muir holds a Certificate in Distance Learning from the University of Washington. She was co-creator and instructor of a course called Internet Safety for Educators, offered through the distance learning programs at both Washington State University and The University of Alaska, Anchorage. Previously, she was an instructor of Technical Writing at Indiana University–Purdue University in Indianapolis. In addition to co-authoring *Our Digital World*, Muir has co-authored *Guidelines for Microsoft Office* for Paradigm Education Solutions and has written a number of technology and business books, including *Distance Learning for Dummies* and *Young Person's Guide to Character Education*, which received the Benjamin Franklin Award for Excellence from the Independent Bookseller's Association.

Denise Seguin
Author, Instructor
Fanshawe College
London, Ontario

Denise Seguin served on the Faculty of Business at Fanshawe College of Applied Arts and Technology in London, Ontario, as a full-time professor, from 1986 until her retirement from full-time teaching in December 2012. She developed curriculum and taught a variety of office technology, software applications, and accounting courses to students in postsecondary Information Technology diploma programs and Continuing Education courses. Seguin served as Program Coordinator for Computer Systems Technician, Computer Systems Technology, Office Administration, and Law Clerk programs and was acting Chair of the School of Information Technology in 2001. Seguin earned her Master of Business Administration, specializing in Technology Management and choosing to take her degree at an online university. In 2016, Seguin returned to teaching courses online on a part-time basis for Fanshawe College. Along with authoring *Seguin's COMPUTER Concepts*, First, Second, and Third Editions, and *Seguin's COMPUTER Applications with Microsoft® Office 2013*, First Edition, *Microsoft® Office 2016*, Second Edition, and with *Microsoft® Office 2019*, she has also authored Paradigm Education Solution's *Microsoft Outlook* 2000 to 2019 editions and co-authored *Our Digital World* First through Fifth, Benchmark Series *Microsoft® Excel®*, 2007 to 2013, Benchmark Series *Microsoft® Access®* 2007 to 2013, Marquee Series *Microsoft® Office*, 2000 to 2013, and *Using Computers in the Medical Office*, 2003 to 2010.

Anita Verno

Professor
Bergen Community College
Paramus, New Jersey

Anita Verno is a professor of Information Technology at Bergen Community College and was IT coordinator/department chair from 2000 to 2010. A founding member of the Computer Science Teachers Association (CSTA) (https://www.csteachers.org/), Verno was the elected "College Faculty Representative" to its Board of Directors and served as curriculum chair from inception until June 2009 when she accepted a position serving on the CTSA Advisory Council. She also assisted in establishing CSTA's Northern New Jersey chapter to serve computing teachers in her home state. Verno served as an associate member of the ACM Committee for Computing Education in Community Colleges (CCECC), where she worked as part of a curriculum project team tasked with identifying the nature and breadth of IT-related programs of study in community colleges. Verno is the former college representative to the CSTA Northern New Jersey chapter, a former president of the Community College Computer Consortium, and a member of the Board of Advisors for the New Jersey Institute of Technology (NJIT) Bachelor of Science in Engineering Technology program. She has over 40 years of professional experience in software design and development, teaching in IT/CS, and development of curricula/degrees for high schools and colleges. Along with co-authoring *Our Digital World*, Verno has co-authored *Guidelines for Microsoft Office* for Paradigm Education Solutions.

Acknowledgments

We would like to thank the following reviewers who have offered valuable comments and suggestions on the content of this program.

Becky Anderson
Zane State College
Zanesville, Ohio

Martin S. Anderson, MBA
BGSU Firelands
Huron, Ohio

Roberta Baber
Fresno City College
Fresno, California

Roxanne Bengelink
Kalamazoo Valley Community College
Kalamazoo, Michigan

Dave Bequette
Butte College
Oroville, California

Shirley Brooks
Holmes Community College
Ridgeland, Mississippi

George Cheng
Hostos Community College
Bronx, New York

Scott Cline
Southwestern Community College
Sylva, North Carolina

James Cutietta
Cuyahoga Community College Metro
  Campus
Cleveland, Ohio

JD Davis
Southwestern College
Chula Vista, California

Alec Fehl
Asheville-Buncombe Technical
  Community College
Asheville, North Carolina

Lisa Giansante, BAS, CMA
Humber College
Toronto, Ontario

Debra Giblin
Mitchell Technical Institute
Mitchell, South Dakota

Glenda Greene
Rowan-Cabarrus Community College
Salisbury, North Carolina

Prosper Hevi
Kankakee Community College
Kankakee, Illinois

Marilyn Hibbert
Salt Lake Community College
Sandy, Utah

Terry Hoffer
City College at Montana State University
Billings
Billings, Montana

Mardi Holliday
Community College of Philadelphia
Philadelphia, Pennsylvania

Stacy Hollins
Florissant Valley Community College
St. Louis, Missouri

Sherry Howard-Spreitzer
Northwestern Michigan College
Traverse City, Michigan

Ly Huong
UCSC Extension
Santa Clara, California

Vincent Kayes
Mount Saint Mary College
Newburgh, New York

David Kern
Whatcom Community College
Bellingham, Washington

Annette Kerwin
College of DuPage
Glen Ellyn, Illinois

Sylvia Knapp
Brunswick Community College
Leland, North Carolina

Paul Koester
Tarrant County College, Northwest
Campus
Fort Worth, Texas

George Kontos, Ed.D
Bowling Green Community College
of Western Kentucky University
Bowling Green, Kentucky

Kathy Lynch
University of Wisconsin–Oshkosh
Oshkosh, Wisconsin

Lana Mason
Wayne Community College
Goldsboro, NC

Lorraine Mastracchio
College of Westchester
White Plains, New York

Dr. Lisa McMillin
East Central Community College
Decatur, Mississippi

Jolene Meyers
Terra Community College
Fremont, Ohio

LeAnn Moreno
Minnesota State College
Southeast Technical
Winona, Minnesota

Larry Morgan
Holmes Community College
Ridgeland, Mississippi

Tammie Munsen
Mitchell Technical Institute
Mitchell, South Dakota

Bonnie Murphy
County College of Morris
Randolph, New Hampshire

Gary Muskin
College of Westchester
White Plains, New York

Patricia Newman
Cuyamaca College
El Cajon, California

Phil Nielson
Salt Lake Community College
Salt Lake City, Utah

Greg Pauley
Moberly Area Community College
Moberly, Missouri

Stacy Peters-Walters
Kilian Community College
Sioux Falls, South Dakota

Pattie Roberts
Mesa Community College
Mesa, Arizona

Vicki Robertson
Southwest Tennessee Community College
Memphis, Tennessee

Wesley Scruggs
Brazosport College
Lake Jackson, Texas

Sue VanLanen
Gwinnett Technical College
Lawrenceville, Georgia

Wilma VanSegbrook
Forest Heights Collegiate Institute
Kitchener, Ontario

Scott Warman
ECPI Technical College
Roanoke, Virginia

Mary Ann Zlotow
College of DuPage
Glen Ellyn, Illinois

Chapter 1

Digital Technologies
Exploring a Wealth of Possibilities

What You'll Accomplish

When you finish this chapter, you'll be able to:

1.1 Recognize the types of digital devices available today.

1.2 Differentiate between the four categories of computers and explain how technological convergence has had an impact on the functions of computers.

1.3 Describe how digital devices are being used and how they provide various career opportunities.

1.4 Explain the information processing cycle and differentiate between data and information.

Why Does It Matter ?

You encounter computers in their various forms, from desktop computers to smartphones to GPS navigation systems, almost every day. In whatever work you do now or are preparing for in the future, understanding the role of computing and the capabilities of computers can help you get ahead. Even a basic level of technical knowledge gives you an edge in the job market, although it is becoming increasingly important to expand your skills beyond the basics to compete for available positions.

 The online course includes additional training and assessment resources.

The world of computers includes several types of computing devices that people use for many different purposes. All digital devices process data and produce information that enriches our personal and professional lives.

Course Content

Take a Survey

Technology in Your Future Video

1.1 Just What Is a Computer?

On the simplest level, a **computer** is an electronic, programmable device that can assemble, process, and store data. An **analog computer** uses mechanical operations to perform calculations, as with an older car speedometer, slide rule, or adding machine. A **digital computer**, such as your laptop, desktop, or tablet, uses symbols that represent data in the form of code. A digital computer has a much higher level of functionality than an analog computing device, including the ability to process words, numbers, images, and sounds.

The combination of digital computing capability and communications has brought us tools such as email and instant messaging and paved the way for new types of computing devices including smartphones, gaming devices, and smart appliances.

Today, many devices beyond the traditional desktop and laptop computers have computing capabilities, including phones and gaming consoles.

1.2 Computers in All Shapes and Sizes

Although there are many devices that have computing capability (which we'll talk about shortly), most of us think of a computer as the desktop, laptop, or tablet we use for work or entertainment. According to a 2017 annual visual networking index forecast by Cisco, there will be four networked devices and connections per person globally by 2021. However, in North America, there will be 13 networked devices and connections per person, up from 8 in 2016.

Take a stroll down the computer aisle of a well-stocked office superstore (or click through its website) and you'll find that there are several kinds of computer models available. First, you'll notice two broad groups of computers: Windows- or Linux-based personal computers (PCs) manufactured by a wide variety of companies such as Hewlett-Packard, Dell, Acer, and Lenovo; and Mac computers, available at Apple stores and other authorized retailers. A third category of computers, Chromebooks, run Google's Chrome operating system for web-based computing and are made by manufacturers such as Samsung and Acer. Manufacturers are also beginning to produce **wearable computers** in the form of glasses, clothing, and watches.

Table 1.1 lists some common computers categorized by size, which usually correlates with **computing power**, or the amount of data that can be processed and the speed at which it can be processed. For example, the computers common in homes and offices range from portable devices to larger units that sit on a desk or table. But the computing world also includes powerful larger computers used in science and medicine.

TABLE 1.1 Examples of Computers by Size

Category	Examples
Larger	Supercomputer, mainframe, server
Mid-sized	Personal computer, desktop, laptop, convertible laptop, ultrabook
Mobile	Tablet, smartphone, digital audio player, e-reader, handheld game console, wearable computer

Supercomputer

The computing power in your laptop today would astound personal computer users of ten years ago. But the fact remains that there are some tasks your personal computer doesn't have the power to handle.

That's where supercomputers and their cousins, mainframe computers (now typically used as servers, which will be discussed in Chapter 6), come in. A **supercomputer** is a computer with the ability to perform trillions of calculations per second. A supercomputer is usually custom-made for a particular use. For example, when a scientist wants to run a computer model of what happens when a star explodes, or a medical researcher has to process the millions of data points from an MRI scan, he or she turns to a supercomputer, because processing the amount of data required could take years on a personal computer. Supercomputer processing power is measured in petaflops. While a **flop** is one floating-point operation per second, a **petaflop** represents a thousand trillion floating-point operations per second.

Supercomputers often function as very large servers in a network. Another model for supercomputers involves **computer clusters**. The moviemaking and computer gaming industries, for example, use **render farms**—clusters of computers joined together with custom-designed connections—to make full-length animated movies and feature-rich games that require high-quality images. These same images would take up to a hundred years to build on a personal computer.

Desktop

A **desktop computer** is a computer whose central processing unit (CPU) is housed in a tower configuration or, in some cases, within the monitor, as with the Apple iMac. Though models vary, a typical desktop computer setup includes a CPU, monitor, keyboard, and mouse. Most desktop models would not be considered portable, though you may be able to move some mini-towers or other compact designs around the house with relative ease.

Laptop

A **laptop**, also known as a *notebook computer*, contains a monitor, keyboard, mouse, CPU, and battery all in one device. Laptops are designed to be portable, although today many are intended as replacements for desktop computers and may come with large monitors that can increase their weight to as much as seven pounds. Other smaller models weigh only three or four pounds and are designed to be taken on the road. While a category of small, easily portable laptops known as **netbooks** became popular in the early 2000s, these lightweight computers contained limited computing power and today have largely been replaced by tablets and **ultrabooks** (very lightweight laptops). Windows 10 PCs with **ARM** are "always-on" PCs. An ARM-powered device operates very similarly to a traditional laptop, but it uses a smaller number of computer instructions and optimized pathways and therefore has extended battery life and improved communication connectivity. The current battery life of an ARM device is 20–22 hours. The Snapdragon computer chip from Qualcomm is an example of a processor that operates with the ARM architecture on mobile devices. Snapdragon is more sophisticated than some ARM-based chips in that it can contain multiple CPU cores, a graphics processing unit, and more.

Laptops weigh less today than they ever have before because most no longer include an optical (CD or DVD) drive. This change is due to the fact that most content can now be accessed online and downloaded from the internet. Another development in the design of laptop computers is the inclusion of touchscreen monitors. Some of these monitors can even be detached from the keyboard and used independently as a tablet, further blurring the lines between the different categories of computers.

The distinctive design of Macs comes from only one source: Apple.

> " The advance of technology is based on making it fit in so that you don't really even notice it, so it's part of everyday life. "
>
> —Bill Gates,
> Co-founder of Microsoft

Tablet

Tablet PCs, which appeared around 2002, were designed to be held like a legal pad and weighed about three pounds. Perhaps inspired by the Star Trek PADD that appeared on the show in 1966, tablet PCs were great for taking to meetings or conferences to make notes either by writing on the screen or by using an onscreen keyboard. Some models also included a traditional keyboard; you could swivel the unit to go from the pad configuration to a more traditional laptop look. These were called *clamshells*.

In early 2010, a new form of tablet appeared on the scene and essentially replaced tablet PCs. The Apple iPad hit the market with high demand and created the benchmark for the new tablet niche. A **tablet** is a portable computing device with a touchscreen interface. A tablet can be used as an e-reader, web browser, and media player, among other functions. Tablets can access a wide variety of apps ranging from games to word processors and spreadsheet software. Several tablets are now available, including the Samsung Galaxy, Google Nexus 10, Amazon Fire, Microsoft Surface, and Apple iPad. The tablet market is expected to continue to grow.

Tablets enable you to use onscreen controls to navigate by touch or with a stylus.

The Convergence of Computing Devices

Technological convergence is a term that describes the tendency of technical devices to take on each other's functions. Today, computers take many forms beyond the traditional desktop or laptop computer. Some are specialized in their functionality; others tackle some big computer tasks despite their compact design. Many combine communications, media, and information processing features in one package. The ability to access many services online rather than loading applications onto a device is making it easier to offer many powerful features in one small package.

One of the most prevalent examples of a **converged device** is the smartphone, which now contains the functionalities of a phone, digital camera, GPS navigation system, and web browsing device, among others. People use smartphones for many tasks (see Figure 1.1), and it is estimated that there are over two billion smartphone users across the globe—an average of more than two smartphones for every nine people. So what are users doing with their smartphones and other converged devices, such as tablets? Which functions are the most popular? Table 1.2 shows the results of a 2017 survey asking people what activities they're using their mobile devices for in comparison to desktop devices. While it's not surprising that mobile devices are being used more for instant messaging and navigation, their use is also outpacing desktop devices for product research, online purchasing, and online banking.

FIGURE 1.1

Technological Convergence

The smartphone is an example of a device that provides a variety of computing features in a small, portable package.

TABLE 1.2

Top 10 Online Activities by Device

Activity	Mobile	PC/Laptop
Visited or used a social network	87%	82%
Used a chat or instant message service or app	83%	43%
Watched a video clip or visited a video-sharing site	74%	68%
Used a map or directions service or app	70%	32%
Visited a news website, app, or service	66%	57%
Searched for a product or service for purchase	63%	59%
Uploaded or shared a photo	57%	33%
Checked the weather	57%	38%
Used an internet banking service	52%	42%
Purchased a product online	51%	49%

Source: https://blog.globalwebindex.net

Course Content
Take the Next Step Activity

1.3 Who Is Using Computers and How?

Computers are used in most businesses to create memos and letters, analyze expenses and sales figures, research and organize data, and communicate with customers. At home, you may have played a computer game, sent email, or watched a movie on your computer. But you may not be aware of how computers are used by people who work in different industries or by people with specialized interests.

Computers Are Everywhere

Although it would probably be easier to give you a list of settings where computers *aren't* being used than to tell you where they are used, the following are some examples of industries in which computers play a major role: government, medicine, publishing, finance, education, arts, law enforcement, the entertainment industry, gaming, and more. **Ubiquitous computing** (also called *embedded technology*) places computing power in your environment to sense information such as temperature and humidity. A refrigerator that tracks its own contents and suggests a shopping list is an example of ubiquitous computing in the home.

Here are a few interesting uses of computers to ponder:

- While you may already be aware that retail websites track your shopping and spending habits online, you may not know that computers are also being used to track your behavior in brick-and-mortar stores. Some stores are now equipped with hidden cameras that contain high-tech facial recognition software. These cameras can be stashed inside mannequins or store displays, and their software can be used for a variety of purposes, from determining a person's age and gender to identifying whether that person is a VIP customer or a known shoplifter. Although this technology can provide retailers with valuable information about their customers, it also raises privacy issues that many shoppers may not be comfortable with.

> “I want the entire smartphone, the entire internet, on my wrist.”
>
> —Steve Wozniak, co-founder of Apple Inc.

- By 2020 it is estimated that one in five cars will be connected to the internet. There will also be approximately 10 million self-driving cars on the streets by 2020.

- The ability to archive electronic images such as CAT scans from medical imaging devices and retrieve them from any location, as well as the use of web conferencing to connect specialists at large urban hospitals with healthcare providers in remote locations, is helping doctors diagnose and treat patients more efficiently.

Automobile manufacturers are experimenting with automated cars, replacing certain driver functions with hands-free computer technology.

- Apple, Nike, and Samsung are using technology to help people live healthier lifestyles through the creation of wearable devices that can track physical activity, heart rate, calories burned, and hours of sleep. These devices interact with apps on smartphones to make it easy for busy users to maintain an active lifestyle.
- Korean retailer Tesco used QR codes to enable customers to order groceries for home delivery from subway stations. Users simply scan a product code with their smartphone from a wall display to add the item to their shopping cart. After they check out, the items will be delivered to their home on the same day. Chapter 2 explains what QR codes are and how to use them.
- Most retailers have installed technology to read RFID chips embedded in debit and credit cards to process transactions. These chips are essentially like computer chips, storing data and initiating actions. This technology replaces the easier to steal magnetic strip and signature method used in older credit cards.
- Location technologies use GPS or Wi-Fi to track people, fleet vehicles such as delivery vehicles, and products and components in transit from one location to another. Location technologies help ensure that employees and customers can find items quickly, manage inventory, enjoy increased security, and have access to items when and where they are needed.

Finding a Career in Computing

Though you are likely to use computers in your work no matter what your profession, there are several computer-specific fields that exist today. These fields are fast-paced and constantly evolving, so new career options come up on a regular basis. Even in bad economic times, employers look to information technology workers to save money with more efficient systems and procedures.

Artificial intelligence (AI), the ability of devices to exhibit intelligence, may open up some new career opportunities in the future as progress is made on developing devices such as autonomous cars.

Today, there are several computer careers you can prepare for, including:

- **Computer engineering (CE).** This field involves the study of computer hardware and software systems and programming how devices interface with each other.
- **Computer science (CS).** People working in this field may design software, solve problems such as computer security threats, or come up with better ways of handling data storage.

> Someday a computer will give a wrong answer to spare someone's feelings and [humans] will have invented artificial intelligence.
>
> —Robert Brault, freelance journalist

- **Information systems (IS).** Those who work with information systems design technology solutions to help companies solve business problems. An IS professional considers who needs what data to get work done and how it can be delivered most efficiently.
- **Information technology (IT).** IT workers make sure the technology infrastructure is in place to support users. They may set up or maintain a network, or recommend the right hardware and software for their companies. IT workers are also responsible for ensuring that devices within a network are able to communicate with one another and share data.
- **Software engineering (SE).** This field involves writing software programs, which might be developed for a software manufacturer to sell to the public or

involve a custom program written for a large organization to use in-house.

A wide variety of careers exists in computer fields, with more being created every day.

- **Web development.** The web provides another group of technology career paths. Programming websites, developing the text and visual content, explaining how clients can use tools such as search engine optimization (SEO) to maximize site traffic, and using social media to promote goods and services are just a few examples of careers in this area.
- **Mobile app development.** Programmers working in mobile app development create programs for devices such as smartphones. These apps have some special requirements that the developers must consider. For example, the apps need to be designed to be usable in a small-screen format and to take advantage of technologies such as GPS to pinpoint the device's location.
- **Database administration.** A worker in this field, called a database administrator (DBA), helps to create and organize databases of information. DBAs make sure that information stored in a database is available and usable. These folks also make sure that data is secure from hackers and unauthorized access.
- **Cybersecurity.** People who work in cybersecurity help to track terrorists, work with law enforcement to capture criminals by pulling information from their computer hard drives, and protect against identity theft and privacy invasion.

According to the *2018 Occupational Outlook Handbook*, computer professionals are in demand, and it's expected that many new jobs will be created in this field between now and 2026. To prepare for a degree in a computer field, you should focus on analytical subjects such as math and science. Employers usually look for a minimum of an associate's degree with a computer science or information technology focus, although many positions require a bachelor's degree or higher. Because computer technology is constantly changing, employers may be more impressed with your résumé if you stay current with changes by obtaining professional certifications. Some employers will pay for or subsidize your professional development; others won't.

> **Course Content**
> Take the Next Step Activity
> Ethics and Technology Blog

1.4 What Is Information Technology?

The Information Technology (IT) department in an organization is in charge of its most valuable asset: information. IT is a department you are likely to come into contact with, whether as an IT employee or as a user of a company computer system, so it's important for you to understand just what information technology is.

According to the former Information Technology Association of America (ITAA), now merged with other organizations to form TechAmerica, information technology is "the study, design, development, implementation, support, or management of computer-based information systems, particularly software applications and computer hardware."

There are a few basic concepts that will help you understand IT. One is the difference between data and information; the other is what's involved in information processing.

Differentiating between Data and Information

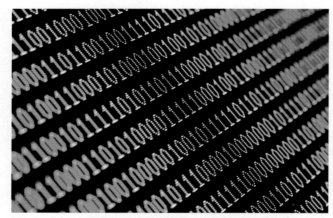

Numerical data may be the basis of information organized in the form of a graph or table.

Say that your job is to predict which new toys will prove most popular next holiday season. You might gather data about popular toys, their prices, and their sales to date. When you take that raw data and organize it into a chart or table that compares sales and prices, it becomes information that you can use to make a business decision.

Computers can take data and turn it into useful information by processing and organizing it. **Data** is what you put into a computer. **Information** is what you can get out of it.

The Information Processing Cycle

What happens between obtaining raw data and getting information based on that data from your computer? That's what the **information processing cycle** is all about. This cycle, shown in Figure 1.2, has four parts: input of data, processing of data, output of information, and storage of data and information.

Input In a typical day, you use a decimal system as you go about your tasks. For example, if you have your car repaired and the bill is $321, you've spent three hundreds, two tens, and one dollar. All this is based on a system of ten possible digits, 0 through 9. Computers, on the other hand, use a **binary system** with two possible values, 0 and 1, called *binary digits*. The term **bit** is a shortening of **binary digits**. Bits are found together in 8-bit collections. Each collection is called a **byte**. Each byte can store one thing like a digit, a special character, or a letter of the alphabet. You'll hear the terms bits and bytes used in descriptions of processors (for example, 64-bit Pentium) and data storage capacity (500 gigabytes memory).

FIGURE 1.2 **The Information Processing Cycle**
There are four parts to the information processing cycle.

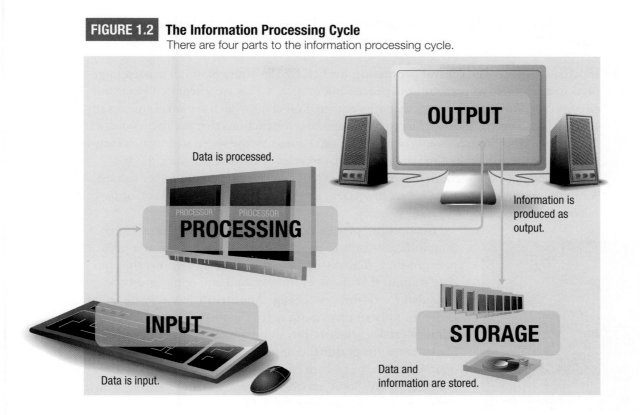

When you **input** data into a computer, it is converted to bits and bytes. Though typing on a keyboard may leap to mind when you think of getting data into your computer, there are actually several methods of inputting data. For example, you might provide your computer with input using a mouse, keyboard, scanner, video game controller, bar code scanner, microphone, camera, or the touchscreen on your smartphone. (You'll learn more about input and output devices in Chapter 3.)

ASCII and Unicode are two computer standards for encoding text characters and symbols for many languages. Essentially, because computers "speak" in numbers, these standards translate the characters of text—letters, punctuation, and other symbols—as well as certain actions such as pressing the Enter key, into numerical terms a computer can understand. **ASCII (American Standard Code for Information Interchange)** is a 7-bit encoding technique. It assigns a number to each of the 128 characters that are used most often in American English. ASCII enables computers to record and display text on your computer monitor. However, ASCII is missing several symbols required to support other languages, such as the German umlaut.

Unicode, however, provides a unique number for all characters, regardless of the operating system, the program using it, or the language. Unicode uses 8-, 16-, or 32-bit characters. This means that Unicode documents can require much more memory than ASCII documents, which are based on a 7-bit system. Today, companies

Devices such as scanners turn existing printed materials into digital data.

such as Apple and Microsoft have gravitated to Unicode which combines several standards, including ASCII, into one central standard.

Processing The **central processing unit (CPU)** in your computer is what interprets instructions and performs the **processing** of data. CPUs are integrated electronic circuits called **microprocessors** that are contained on chips, which are small squares of silicon. These microprocessors accept programming instructions that tell them what to do with the data they receive. Processors are often rated by the speed with which they can process the data, measured in hertz (Hz), or cycles of current per second. One hertz is one cycle per second. One megahertz (MHz) is one million cycles per second. A notebook computer might have a processing speed of two gigahertz (GHz), or two billion cycles per second.

Processor speed has increased dramatically in recent years.

In addition, processors can be 32-bit or 64-bit, which is an indicator of how much data they can handle at a given point in time. The 64-bit processors are more powerful, but they require that the operating system and applications be designed for 64-bit processing. Another differentiating factor is the number of processing cores the microprocessor contains, because each core can simultaneously read and execute instructions. In personal computing, there are dual-core (2), quad-core (4), hexa-core (6), and octo-core (8) processors available today. Some processors, such as Intel's Atom, which draws less power, have specialized features that make them a good match for mobile computing.

While a computer is processing data, it temporarily stores both the data and instructions from the CPU in **computer memory** in the form of **random access memory (RAM)**. There is a constant exchange of information between the CPU and RAM during processing. When you turn your computer off, the data temporarily stored in RAM disappears; therefore RAM is also referred to as **volatile memory**. Think of RAM as similar to a shopping cart—while you're shopping, you can temporarily place items in the cart. If you walk out of the store without buying the items, somebody will empty the cart, causing the items to "disappear."

There are two basic types of RAM: **dynamic RAM (DRAM)** and **static RAM (SRAM)**. Over time, these types have been improved and the names have varied, but the difference in the way they work remains the same. DRAM is the type of memory most commonly found in computers. When we buy a computer or talk about our computer's memory as RAM, we are usually talking about DRAM. DRAM works quickly and is compact and affordable; however, it must have electricity to exist. It is so fragile that the data held in DRAM must constantly be refreshed.

SRAM is a different type of memory. SRAM is about five times as fast as DRAM, and, while dependent upon electricity, it does not require constant refreshing. For this reason, SRAM is much more expensive than DRAM and is typically used only in relatively small amounts. It is built directly into the CPU in the form of cache memory.

Cache memory is an area of computer memory between RAM and the processor that serves as a holding area for the most frequently used data. As illustrated in Figure 1.3, your processor checks the cache memory first, because it is located on or near the microprocessor chip and therefore quicker to access. This procedure saves the processor from having to troll through the entire RAM holding area to find what it needs.

FIGURE 1.3 Cache Memory

Cache memory quickens computer processing time.

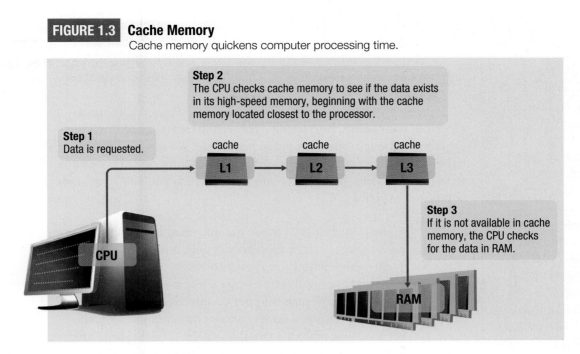

Step 1
Data is requested.

Step 2
The CPU checks cache memory to see if the data exists in its high-speed memory, beginning with the cache memory located closest to the processor.

cache L1 → cache L2 → cache L3

Step 3
If it is not available in cache memory, the CPU checks for the data in RAM.

CPU

RAM

A significant improvement in DRAM access speed came with the invention of an updated version of DRAM called **synchronous DRAM (SDRAM)**. Most modern computer memory is some variation of SDRAM, including DDR-SDRAM, DDR2-SDRAM, and DDR3-SDRAM. Performance has steadily improved with each successive generation.

During processing, the **machine cycle** defines what goes on in your computer when an instruction is sent to it. The machine cycle involves four basic operations:

1. The computer fetches an instruction from memory and places it in the **instruction register**.
2. The instruction is decoded so the computer knows what it's being asked to do.
3. The computer executes the instruction.
4. The computer stores the result of the operation in its main memory for later use.

A program might initiate a series of machine cycles when you press a function key on your keyboard, make a choice in a menu, or click a button on a toolbar.

Output Once you've put data into your computer, you'll typically want to see that data in some form. **Output** is the information that results from computer processing. Output might include the information you view on your monitor, a printed hard copy of a document, or an X-ray produced by a medical imaging computer. Your mobile device's screen can also provide output.

A monitor can display your computer's output at various resolutions. **Resolution** refers to the number of **pixels** (short for picture elements, which relates to the number of dots of color) used to generate an

Monitors come in a variety of sizes and types. This 27-inch widescreen monitor provides a large-screen view of computer programs and data and is especially beneficial for users who work with detailed drawings.

image on your screen. The higher the number of pixels, the more clearly defined the image.

Speakers are another form of output device. You might receive output of audio files through your computer or smartphone speakers, for example.

Printers are another method of output. They create a hard-copy record of your computing results. Though there have been predictions of a paperless office for years, people still seem to like paper copies of documents to keep records or make notes on. Many printer types are available, such as laser and inkjet, which you will learn more about in Chapter 3.

Storage **Storage** is a key part of the information processing cycle. If you have ever spent hours creating a document only to discover that your work was somehow lost, you know how important storage is. Your computer temporarily stores data while it runs processes, but that data is lost when you turn off your computer. You use a permanent storage device, such as a flash drive or your computer's hard drive, to save a version of your work that will be available long after you have shut off your computer. Early permanent storage media included paper punch cards, followed by floppy plastic disks and then floppy disks in hard cases. The most common long-term storage medium is your computer's internal hard drive, which is a metallic disk that uses magnetic or flash technology to store data. You can also store data on removable media (a flash drive or DVD, for example) so you can retrieve it using any computer. Today, an increasingly popular storage medium is **cloud storage**, which involves storing information on the web. Users can access the stored information from any computer with an internet connection by entering a username and password. A popular cloud storage service is Microsoft's OneDrive. OneDrive is free, providing users with 5 GB of free space and additional space up to 50 GB for a fee. Any type of file may be uploaded and set to private, shared, or public. Other popular cloud storage services are Google Drive, Dropbox, iCloud, Box, and ADrive.

Flash drives (also called *USB sticks*) provide lots of storage in a small package.

External hard drives provide an easy way to get extra storage capacity.

The basic unit of storage for data is the **file**, which might contain a report, spreadsheet, or picture, for example. Storage capacity is measured in kilobytes (approximately 1,000 bytes), megabytes (approximately 1 million bytes), gigabytes (approximately 1 billion bytes) and terabytes (approximately 1 trillion bytes).

Storage devices include your computer's hard disk, CDs or DVDs (also called *optical discs*), USB or flash drives (thumb-sized cartridges that slot into a USB port on your computer), and external hard drives (small boxes that connect to your computer via a cable in a USB port). Future forms of computer memory and storage may be made of conductive gels, making them suitable to use in conditions that would destroy current types of memory and storage, such as wet environments.

> **Course Content**
> Take the Next Step Activities

Review and Assessment

Course Content
The online course includes additional
training and assessment resources.

Summing Up

1.1 Just What Is a Computer?

On the simplest level, a **computer** is a programmable device that can assemble, process, and store data. An **analog computer** uses mechanical operations to perform calculations, as with an older car speedometer or a slide rule. A **digital computer** uses symbols that represent data in the form of code.

The combination of digital computing capability with communications has brought us tools such as email and instant messaging and new types of computing devices including mobile phones, gaming devices, and GPS navigation systems.

1.2 Computers in All Shapes and Sizes

Computers fall into three broad groups based on operating system and manufacturer: Windows- or Linux-based personal computers (PCs) manufactured by a wide variety of companies such as Hewlett-Packard, Dell, Acer, and Lenovo; Mac computers available only through Apple; and Chromebooks, which use Google's Chrome OS. Manufacturers are also beginning to produce **wearable computers** in the form of glasses, clothing, and watches, but this category of computers is still emerging.

Computing power refers to the amount of data that can be processed and the speed at which it can be processed. A **supercomputer** is a computer with the ability to perform trillions of calculations per second and is usually custom-made for a particular use.

A **desktop computer** includes a central processing unit (CPU) housed in a tower configuration or, in some cases, within the monitor, as with the Apple iMac.

Also known as *notebook computers*, **laptops** are usually portable, though today many are intended as desktop replacements and therefore may come with a large monitor and weigh as much as seven pounds. **Netbooks**, which became popular in the early 2000s, were lightweight and portable but contained limited computing power. Today they have largely been replaced by tablets and **ultrabooks** (very lightweight laptops). An **ARM**-powered device operates very similarly to a traditional laptop but has battery life of 20–22 hours.

Tablets are manufactured by several companies and enable you to navigate by touching easy-to-use onscreen tools. Tablets run several newer operating systems such as Apple's iOS and Android.

Technological convergence is a term that describes the tendency of technical devices to take on each other's functions, resulting in communications, media, and information computing features being combined into one package. So-called **converged devices** include smartphones, which now contain the functionalities of a phone, digital camera, GPS navigation system, and web browser, among others.

1.3 Who Is Using Computers and How?

Computers are used in government, medicine, publishing, finance, education, arts, law enforcement, social settings, the entertainment industry, gaming, and more.

Though you are likely to use computers in your work no matter what your profession, there are several computer-specific fields that exist today. **Computer engineering (CE)** involves studying computer hardware and software systems and programming devices to interface with each other. **Computer science (CS)** workers may design software, solve problems such as computer security threats, or come up with better ways of handling data storage. **Information systems (IS)** workers identify the kinds of data company employees need and design technology systems to solve business problems. **Information technology (IT)** workers make sure the technology infrastructure is in place to support users. They may set up or maintain a network or recommend the right hardware and software for their companies. **Software engineering (SE)** involves writing software programs that a software manufacturer sells to the public, or custom programs that a large organization uses in-house. There are also a number of **web development**—focused career tracks, as well as jobs for mobile app developers, database administrators (DBAs), and **cybersecurity** analysts.

1.4 What Is Information Technology?

According to the former Information Technology Association of America (ITAA), now part of TechAmerica, information technology is "the study, design, development, implementation, support or management of computer-based information systems, particularly software applications and computer hardware."

Computers take **data** and turn it into useful **information** by processing and organizing it. For example, if you take raw data about sales and prices and put it into a chart, it becomes information that you can use to make a business decision.

The **information processing cycle** has four parts: input of data, processing of data, output of information, and storage of data and information. Computers use a **binary system** of data based on two possible values, 0 and 1, called **binary digits** or **bits.** Bits are found together in 8-bit collections. Each collection is called a **byte**. Each byte can store one thing like a digit, a special character, or a letter of the alphabet.

You provide your computer with **input** using devices such as a mouse, keyboard, scanner, game controller, bar code scanner, or the number pad on your smartphone. Computer standards for encoding text characters and symbols, such as **ASCII (American Standard Code for Information Interchange)** and **Unicode**, translate the characters of text and actions into numerical terms a computer can understand.

The **central processing unit (CPU)** in your computer interprets instructions and performs the **processing** of data. CPUs are made up of integrated circuits called **microprocessors** contained on chips. Processors are often rated by the speed with which they can process data, measured in hertz (Hz), megahertz (MHz), or gigahertz (GHz). While 64-bit processors generally can handle more data at once than 32-bit processors, they also require matching 64-bit software. The number of cores indicates how many simultaneous program instructions the processor can execute, with dual-core (2), quad-core (4), hexa-core (6), and octo-core (8) processors available today. Some processors, such as Intel's Atom, which draws less power, have specialized features that make them a good match for mobile computing.

While a computer is processing data, it temporarily stores both the data and instructions from the CPU in **computer memory** in the form of **random access memory (RAM)**, also called **volatile memory**. When you turn off your computer, the data temporarily stored in RAM disappears. There are two types of RAM: **dynamic RAM (DRAM)** and **static RAM (SRAM)**. A significant improvement in DRAM access speed came with the invention of an updated version of DRAM called

synchronous DRAM (SDRAM). Most modern computer memory is some variation of SDRAM.

There is also an area of computer memory between RAM and the processor called *cache memory.* **Cache memory** is a holding area for the most frequently used data. During processing, the **machine cycle** defines what goes on in your computer when it receives an instruction.

Output is the information that results from computer processing. Output can include the information you view on your monitor or a printed hard copy of a document, among others.

Storage is a key part of the information processing cycle. Your computer temporarily stores data while it runs processes, but that temporarily stored data is lost when you turn off your computer. You use a permanent storage device to save a copy of your data or information that's available long after you have shut off your computer. The basic unit of storage for data is the **file**. Storage devices include your computer's hard disk, CDs or DVDs (also called *optical discs*), USB or flash drives, and external hard drives. **Cloud storage** involves storing information on the web. Future memory and storage technologies might use conductive gels that can stand up in wet environments.

Terms to Know

1.1 Just What Is a Computer?

computer, 2
analog computer, 2
digital computer, 2

1.2 Computers in All Shapes and Sizes

wearable computer, 3
computing power, 3
supercomputer, 3
flop, 3
petaflop, 3
computer cluster, 4
render farm, 4
desktop computer, 4
laptop, 4
netbook, 4
ultrabook, 4
ARM, 4
tablet, 5
technological convergence, 5
converged device, 5

1.3 Who Is Using Computers and How?

ubiquitous computing, 7
artificial intelligence (AI), 8
computer engineering (CE), 8
computer science (CS), 8
information systems (IS), 8
information technology (IT), 8
software engineering (SE), 8
web development, 9
mobile app development, 9
database administration, 9
cybersecurity, 9

1.4 What Is Information Technology?

data, 10
information, 10
information processing cycle, 10
binary system, 10
bit, 10
binary digits, 10
byte, 10
input, 11
American Standard Code for Information Interchange (ASCII), 11

Our Digital World

Projects

Check with your instructor for the preferred method to submit completed work.

Computers Then and Now

Project 1A

As a computer user, you may be curious about the people behind the development of computers and the devices you connect to them. To gain a better understanding of these people, research on the internet and prepare a pictorial timeline of the work of these innovators along with their contributions. Your pictorial timeline should include 10 to 15 innovators who contributed to the development of computer hardware. The following questions, along with the names of the innovators you choose, will provide a head start for your timeline.

1. A slide rule is often associated with the early days of information tools. Who is associated with its invention?
2. Who is known as the "Father of Computers"?
3. Who invented Z series computers?
4. The first fully electronic digital computer, known as the Atanasoff-Berry Computer (ABC), was invented by a professor and his student. What were their names?
5. Who invented the input device known as the mouse?

Be prepared to present your completed timeline according to your instructor's directions.

Project 1B

Because of the prevalence of computers in today's society, it's essentially guaranteed that you and your classmates will work in jobs that require the use of some type of computing device. Select two different careers that are of interest to you. Research the ways in which computers and technology are used in performing these jobs by addressing the following questions:

1. What aspects of the job require the use of a computing device?
2. In what ways does using a computer support efficiency, effectiveness, and productivity in this position?
3. How have computers changed the way the job is performed over the last 20 years?
4. What types of advances in computer technology do you think might affect this job in the future?

Prepare a presentation summarizing the information you discover. Be prepared to share the presentation with the class.

Making Computer-Related Recommendations

Project 1C

As the IT manager for a large physician's practice you have been asked by the practice manager to determine what type of computing devices to provide to the physicians in the practice. The physicians will use these devices both for their own work and to provide healthcare information and services to patients. Research the capabilities of smartphones, tablets, and laptops in a healthcare setting. Using the information you have found, prepare a table that addresses the following questions:

1. What is the approximate cost of each category of device?
2. What features does each type of device offer?
3. What components could be added to each device to improve functionality?
4. How can each device uniquely improve how the physicians deliver healthcare services?

Use the information from your table to determine which type of computing device would be the best option for the physicians in the practice. Prepare a report making a recommendation of the type and model of device the practice should provide and include details to support your recommendation. Submit the report to your instructor.

Project 1D TEAM

The owners of a toy manufacturing company believe that new computer input and output devices will result in improved worker efficiency. Your IT team has been asked to study the issue and make a recommendation. To accomplish this project, you and your teammates should complete the following tasks:

1. Prepare a list of all possible input/output devices in general use.
2. Describe what type of training each input/output device requires.
3. Identify which company departments (such as marketing, sales, finance, manufacturing, customer service) might use a specific input/output device.
4. List the cost of each device, assuming that mice and keyboards do not need to be purchased. Include your sources.
5. Recommend three input/output devices that should be purchased.

Prepare a table that displays your research findings, and attach a summary of your written recommendation.

How to Avoid Computer-Related Injuries

Project 1E

Substantial evidence indicates that individuals who frequently use computers at work are at risk for computer-related injuries. As you start your new data entry job, you want to ensure that you don't become another statistic. Research the topic of work-related injuries for computer users. Prepare a blog post that lists common injuries, the causes of these injuries, and suggestions for creating a safe working environment.

Project 1F TEAM

As members of the human resources department, your team has been assigned to investigate an increase in work-related employee injuries. Specifically, employees who work at computer workstations have reported chronic musculoskeletal problems. To help these employees, your team will prepare a slide presentation on how to create an ergonomic workstation. Your presentation should address chair requirements, including height adjustments and lumbar support, monitor and keyboard locations and settings, ergonomic accessories, and training. Be prepared to share your presentation with the class.

Storage and Processing

Project 1G

You recently left your flash drive in the college computer lab and lost all of your assignments. You now have to redo all your work. You have heard of cloud storage services such as OneDrive, Google Drive, and Dropbox from your friends. You decide to research the different types of cloud storage to back up your assignments. Using the internet, research four different cloud storage services and prepare a table that includes the following information:

- Web address of the service
- Image of the cloud service website
- Storage fee, either by month or year (or indicate if free)
- Storage capacity
- Speed of uploads and downloads
- Available support services

Include with your table a written summary of your research, and state which cloud storage service will best meet your needs. Be sure to document your resources.

Project 1H TEAM

Read or listen to the "History of the Computer Processors" article at https://ODW5 .ParadigmEducation.com/ProcessorHistory. Prepare a presentation about how the processor has evolved and improved the performance of computing devices.

Wiki Project—Backing Up Data

Project 1I

Interview one person who works at your college or a local business to find out how and why they back up data. If possible, pursue interesting industries that might have unique data needs such as nuclear power facilities, hospitals, or the stock exchange. Post your interview in podcast or transcript form on the class wiki. Be sure to include references for verification of content.

Project 1J TEAM

Natural disasters often occur without notice. For example, in 2011, a tornado damaged St. John's Hospital in Joplin, MO. Fortunately, the hospital had begun to use electronic health records. Even though the hospital was extensively damaged, the patient records were still accessible because the hospital had a disaster recovery plan that enabled employees to access the data and information. In class, break into small teams and design a disaster recovery plan to be used in a particular industry such as health care or telecommunications. The disaster recovery plan might include backup methods and comments about why these methods are important for that industry. Every team should post the information it gathered about disaster recovery plans on the class wiki. If two teams post information about the same industry, they should combine the entries. In addition to posting an entry, add a comment or more information to other teams' posts. When you post or edit content, make sure you include a notation with the page that includes your team name and the date you posted or edited the content.

Choosing a Computing Device

Project 1K

You are the Director of IT for a two-year community college. The dean of the business school has informed you that all incoming business students will be required to purchase a computing device to be used in the BYOD (bring your own device) pilot project. The dean wants to provide a list of recommended devices to the students. Research computing device options, focusing on current tablets. Prepare a table or spreadsheet listing the tablet specifications and costs. Include any upgrades or additional components (along with their costs) that would enhance the BYOD system. In a memo addressed to the dean, state your computing device recommendations and include your table or spreadsheet showing your cost analysis. In addition, provide a list of sources used in your research.

Project 1L

You're ready to purchase a new smartphone to take advantage of newer technologies. There are many choices. Using the internet, research three different smartphones from manufacturers such as Samsung, Apple, or Nokia. Prepare a presentation that includes an image of each smartphone and discuss the operating system, weight, screen size, dimensions, camera specs, screen resolution, battery life, features, and carriers. Explain what you use a smartphone for and make a decision based on the features you use most. Submit your presentation to your instructor. Be prepared to share your research with the class.

Class Conversations

Topic 1A   Do computers cause us to work less efficiently?

Computers have changed the way we work. Most employees would say that computers have helped them become more effective and efficient in the workplace. Some think that computers just create more work and cause problems when they crash or slow down. Discuss a career or place of employment where you think computers have actually decreased effectiveness and efficiency. Why do you think this is the case? Be prepared to support your position.

Topic 1B   What will computing in the future look like?

Technology changes rapidly. Since the first personal computer appeared on the cover of *Popular Electronics* in 1975, the computer has changed in performance, cost, size, and capabilities. The personal computer has morphed from a desktop to a handheld device. Consider what you think the computer of the future will be like. Discuss what the computer might look like, as well as its functionality, capacity, and performance.

Topic 1C What is the impact of going overseas for technology services?

Computers and technology have enabled outsourcing by large organizations for many years. Now, small businesses often need workers for a one-time project such as website design or software development, so they are using online job marketplaces that contract with skilled foreign workers at lower costs. Discuss the advantages and disadvantages of small businesses outsourcing technology work to other countries. Discuss how this will have an impact on the US and Canadian economies and small business job markets in the future.

The Internet
Gateway to a World of Resources

What You'll Accomplish

When you finish this chapter, you'll be able to:

2.1 Describe how the internet and the web have changed the ways in which people interact with each other.

2.2 Describe the internet and the web in terms of the internet's infrastructure and the web's phases of evolution.

2.3 Select the services, equipment, and software you need to connect to and browse the internet.

2.4 Describe the organization of web pages and how they are accessed.

2.5 Distinguish between appropriate and inappropriate use of intellectual property and copyright as they apply to the web.

2.6 Differentiate among the three types of e-commerce.

2.7 Compare various internet services and applications such as email, instant messaging, and audio and video conferencing.

Why Does It Matter ?

From protecting our health to electing our officials, the internet has become an indispensable part of our lives. Online caring site CaringBridge is visited by over 500,000 people a day to get updates about and support loved ones who are dealing with healthcare challenges. A study by IBM indicated that 80% of the students surveyed "anticipate running into new technology that they will have to adapt to and learn upon entering the workforce." The internet can be used for this purpose. By the time President Obama was starting his second term in office in 2013, he had already supported changes to better control digital data, appointed the first federal chief technology and chief information officers in the country, and made the position of the White House cybersecurity coordinator more prominent. The Trump administration has called for making electronic health records easier for patients to use. Understanding what the internet is, how it works, and the dynamics of its growth will help you succeed personally and professionally in our digital age.

 The online course includes additional training and assessment resources.

The internet has changed the way we live, work, and play. From communicating to searching for information to socializing to conducting business, a vast web of resources and activities is at our fingertips 24/7.

Course Content

Take a Survey

Technology in Your Future Video

2.1 The World Goes Online

Many people, for much of the day, use the internet for a wide variety of activities. People learn, work, play, and connect with others using several types of devices, such as computers, tablets, smartphones, gaming devices, and information kiosks. We can even **sync** our own electronic devices with one another to keep our data up to date. This enables us to do things like view our calendar and contacts and post to social media and other sites from any internet-connected device.

Your smartphone is a tiny computer with enormous potential.

According to the Pew Internet Project, as of January 2018, 95% of American adults owned a cell phone, 77% owned a smartphone, 53% owned a tablet, and more than 22% owned an e-reader. As of May 2017, more than half of US households had one or more cell phones but no landline. As of 2018, according to Pew Research, nearly 75% of adults and 94% of 18–24 year olds used social media. Of all adult internet users, 54% posted original photos or videos that they themselves had taken, and 47% took photos or videos they found online and reposted them on sites designed for sharing with many other users. All these activities performed using various devices take advantage of the internet to do things that would have seemed miraculous only a dozen years ago.

Cutting-Edge Internet

Activities such as shopping, communicating, and researching online are probably familiar to you, but some other uses of the internet may be new to you. For example:

- Twitter is being used by politicians and celebrities to share their political and personal views and to make us aware of life-changing events like births, marriages, and divorces.

- The world of retail shopping is changing as stores embrace technology in innovative ways. Retail stores like Amazon Go in Seattle are providing apps that allow self-checkout and payment via smartphone with no need to wait in line.

Amazon Go stores provide a checkout-free experience for buyers with an Amazon account and a current smartphone.

- Online shopping is growing. Cyber Monday in 2017 was the largest online shopping day in US history with $6.59 billion in online sales. The trend toward online shopping is increasing worldwide. Retail e-commerce sales in 2016 were 1.86 trillion US dollars. In 2021, they are expected to be 4.79 trillion.
- Video and audio technology is making it possible for patients in rural areas to visit a medical specialist without leaving their homes.
- Voice-activated speakers, or **smart speakers**, such as Amazon Echo and Google Home are simplifying interaction with computers, making them easily accessible for simple tasks (see Figure 2.1). Using a smart speaker, you can play your favorite music, obtain answers, and even control parts of your home via a built-in "home assistant" feature.
- Thirty-nine percent of households contain at least one streaming media device, such as Apple TV, Roku, or Google Chromecast.

FIGURE 2.1 **Speaking Your Mind**
Smart speakers such as Amazon Echo are making it easier to get answers and help with no keyboard involved.

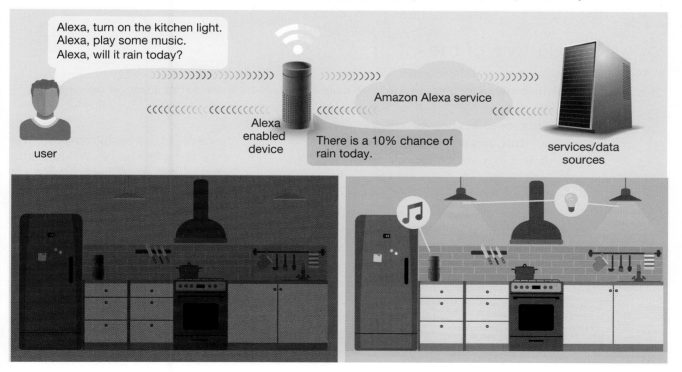

- Wearable devices and smart appliances are helping to make computers ubiquitous.
- Airline pilots practice landing a plane on water with web-based flight simulators.
- A **QR (quick response) code** provides a shortcut you can use to go to a website using your smartphone. Rather than entering the web address, you use your phone (with a reader application installed) to scan this 2-D code and let it connect your phone to the site. QR codes are a marketing tool used to bring potential clients or customers to a target website. A school library can use a QR code to direct a student to a website to search for a particular book.
- More people make purchasing decisions based on information gleaned from social media, forcing today's corporations to include social media as a regular and vital part of every marketing plan.
- Cloud computing lets you access your files and programs from anywhere in the world via the internet. Because the files and programs can be accessed using the web, they are always available.

Many colleges are phasing out computer labs and requiring that freshmen have a laptop computer to complete coursework and communicate with instructors.

- Your car has become a computer. Driver assistance technologies have been incorporated into some newer vehicles to help keep us safe. Changes in the roadway are detected using cameras, sensors, and radar and processed using computer software. Together, these technologies may help a driver stay in a lane, avoid backing up into another vehicle, and apply the brakes if necessary. Driverless or autonomous cars are being tested for safety and may be the norm in the future.

What remarkable uses of the internet are you aware of?

The Future of the Internet

Some current uses of the internet are amazing, and in the future, this technology will only become more ingrained in your daily life. Consider some recent advances in technology:

- Technology has become a bigger part of our daily lives with the incorporation of devices that process spoken commands such as Amazon Echo and Google Home. These devices sync to your calendar, contacts, and the internet. They use artificial intelligence to interpret your spoken commands and make appropriate responses. Communication among these devices will eventually make them indispensable.

> " You can't just ask customers what they want and then try to give that to them. By the time you get it built, they'll want something new. "
>
> —Steve Jobs,
> former CEO of Apple

- In 2017, regulations implementing net neutrality were repealed. Net neutrality specifies that all internet traffic should be treated equally. With the repeal of net neutrality, it is possible for an internet service provider (ISP) to give higher priority and more bandwidth to content from certain companies who pay a fee and restrict bandwidth for others.

- Nanotechnology-enabled robots can be injected into your body, move around to deliver drugs and monitor vital signs, and then send information to your doctor over a wireless internet connection.
- Smartphones can connect to your smart home appliances via the internet to set the temperature on your thermostat or turn on the coffee pot when you wake up in the morning.
- In the future, ambient computing using devices that are sensitive and respond to humans will become increasingly popular. For example, when your car pulls into your driveway, the garage door opens, the front door unlocks, and the lights inside the house turn on. The temperature is already set, and music starts to play based on your mood. You're reminded of calls you should make and your mom's birthday the next day.

In fact, what was once science fiction is becoming reality through technologies that are being thought of or developed today. Nobody imagined texting or video streaming over YouTube twenty years ago, yet these types of significant additions to our culture explode on the scene rapidly. Tomorrow is anybody's guess—will you be there to adopt or even introduce the next great online innovation?

2.2 What Are the Internet and World Wide Web?

Together the internet and World Wide Web (better known as simply the web) have made it possible for the world to connect and communicate in remarkable ways. But what role does each of them play?

What Is the Internet?

The **internet** is the world's largest computer network, made up of many smaller networks that provide a physical infrastructure that allows us to share resources and communicate with others around the world. It is possible to see, touch, and feel elements of the internet, because it's made up of hardware such as servers, routers, switches, transmission lines, and towers that store and transmit vast amounts of data (Figure 2.2).

The internet is the pathway on which data, sounds, and images flow from person to person around the globe.

What Is the Web?

If the internet is a pathway for information, the web is one type of content that travels along that path. The **web** is a body of content that is available as web pages. The pages are stored on internet servers around the world. A **web page** may contain text, images, interactive animations, games, music, and more. Several web pages may make up a single **website** (Figure 2.3).

> "I must confess that I've never trusted the web. Where does it live? How do I hold it personally responsible? And is it male or female? In other words, can I challenge it to a fight?"
>
> —Stephen Colbert, Comedian

The documents, images, and other information that you choose to put on the web are the things you want other people to see. You can also choose whether you want everyone in the world to view your content or limit access to a few close friends.

FIGURE 2.2
The Infrastructure of the Internet
Data moves across the internet by traveling through a collection of physical objects.

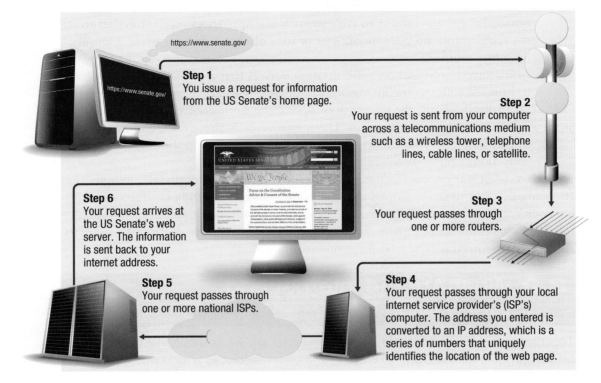

https://www.senate.gov/

Step 1
You issue a request for information from the US Senate's home page.

Step 2
Your request is sent from your computer across a telecommunications medium such as a wireless tower, telephone lines, cable lines, or satellite.

Step 6
Your request arrives at the US Senate's web server. The information is sent back to your internet address.

Step 3
Your request passes through one or more routers.

Step 5
Your request passes through one or more national ISPs.

Step 4
Your request passes through your local internet service provider's (ISP's) computer. The address you entered is converted to an IP address, which is a series of numbers that uniquely identifies the location of the web page.

FIGURE 2.3
A Website
A single website is made up of one or more web pages. You use a web browser, a type of software, to find and view a website.

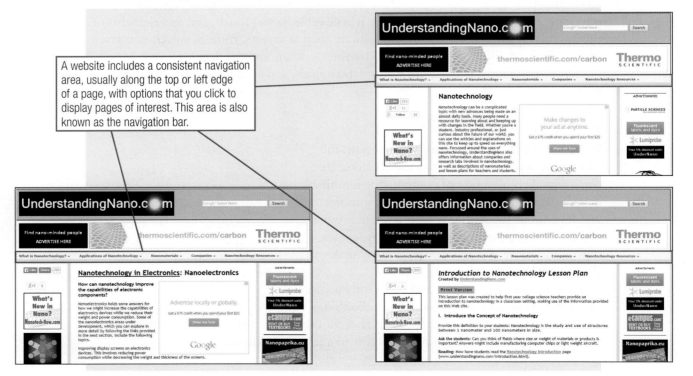

A website includes a consistent navigation area, usually along the top or left edge of a page, with options that you click to display pages of interest. This area is also known as the navigation bar.

Because anyone can publish just about anything to the web, it has become very popular very fast, and has grown with little oversight or regulation. Many people feel that what makes the web appealing is that it is a largely unregulated environment. However, that very lack of regulation and supervision means that some people feel free to abuse others or commit crimes using online tools and sites. The web is therefore sometimes referred to as the wild, wild West.

Although certain countries have regulations and laws in place regarding the web, it is hard to enforce them across borders and cultures. For example, many forms of online gambling are illegal in the US, but it's easy for US citizens to gamble online using off-shore hosted online casinos. Still, many people feel that the benefits of freedom of expression and sharing of ideas and information on the web far outweigh the problems that result from the lack of regulation and control.

The different phases of the web are not like software releases. Rather, they represent new trends in online usage and technologies brought about both by the way people use the internet and technologies that enable new activities in the online world. Table 2.1 lists the three phases of the web and describes the significant capabilities of each phase.

TABLE 2.1   Three Phases of the Web

Phase	Also Known As	User Capabilities
Web 1.0	Static web	Read content but can't interact with anything
Web 2.0	Interactive web	Read content, interact with content, write content, collaborate
Web 3.0	Semantic web	Read content, interact with content, write content, collaborate, access connected content, use devices that can talk to one another to gather information and present useful information

Web 2.0 We have already experienced the first shift in online usage. This shift was the start of the second phase of the web, referred to as Web 2.0. **Web 2.0** was thought of as an interactive and social web that facilitated collaboration among people.

The concept of Web 2.0 came out of the "dot-com collapse" of 2001, when many online businesses failed. Internet users appeared to be unwilling to pay fees to use online services such as email and news and sought a more collaborative experience. New applications and websites began popping up with surprising regularity, marking a turning point for the web, according to an article by Tim O'Reilly, founder of O'Reilly Publishing.

💻 Computers in Your Career

The internet and web have made available a wide variety of careers that didn't exist 10 years ago, and not all are high-tech in nature. For example, consider jobs such as web content writer, internet law expert, and online trainer. If you are more technically inclined, a variety of developer jobs allow you to work with tasks from programming an e-commerce shopping cart to creating environments in virtual reality worlds.

Over time, Web 2.0 has become associated with interactive web services such as Wikipedia and Facebook that provide users with a way to collaborate by sharing and exchanging ideas and adding or editing content. The premise of this generation of web usage is that people are not passively viewing information but are also interacting with and helping to create content.

Web 3.0 **Web 3.0**, which has already begun and will evolve over time, is seen as the next phase in online usage. Web 3.0 will host collaborative content that is connected in meaningful ways. For example, while in the past you could post a photo from your vacation online and also keep an online calendar with your trip itinerary, in the world of Web 3.0 your photos appear within the calendar on the date and time you took them.

Web 3.0 introduces the ever-present web—the merging of computers and mobile devices as a source for music, movies, and more. This puts the internet at the center of both our work lives and our leisure time.

Web 3.0 makes people's online lives easier and more intuitive as smarter applications such as better search functions connect users with the information they seek. The internet will evolve into an artificial intelligence that understands context rather than simply comparing keywords. Artificial intelligence permits virtual assistants to interpret natural language to understand what you want. This is one reason that Web 3.0 is also called the **Semantic Web**, appropriately named because semantics is the study of meaning in language.

It is thought that Web 3.0 will help grow the **Internet of Things (IoT)**—connecting devices that contain embedded computing chips to enable data transfer over a network without human involvement.

In Web 3.0, machines can talk to machines. For example, TiVo searches the internet and gathers information from various websites about programs that you might like to watch. The Semantic Web makes it possible for websites to "understand" the relationships between elements of web content. An intelligent agent, or search program, may even be able to update and modify your documents on the fly based on information you read on the internet. For this to come about, we have to find common formats so that data can be integrated, rather than just exchanged.

Conducting a search on your computer can return results that include a list of related files, applications, or features stored on your hard drive, plus websites, music, or images from the web. Microsoft's Bing search service finds information and

Computers in Your Career

Huge volumes of patient data create a record-keeping challenge in the medical field. Medical informatics professionals work to tame the mountains of data by designing medical information systems that make it easier to enter, update, and retrieve medical data. This drive toward online medical records was given a boost when the US government mandated digital record keeping from the healthcare industry as of January 2014. Today, in addition to making data retrieval easier, expert systems have been created to present and summarize information in a way that assists medical professionals in diagnosing and treating illnesses. Portals are available that provide patients with access to their medical data. These portals must be maintained and secured by technical professionals.

provides recommendations to help you make choices and decisions for shopping, travel, and more. Stephen Wolfram's WolframAlpha search engine lets you ask questions in natural human language rather than entering keywords. Instead of simply searching for matches, this engine computes answers to your questions.

> **Course Content**
> Take the Next Step Activities
> Ethics and Technology Blogs

2.3 Joining the Digital World

You may have been going online for many years, and it may seem as natural as turning on a television set, or you may have only skimmed the surface of the internet, which came into the public mainstream less than 20 years ago.

Whichever the case, you are probably aware that today, the internet is truly a global phenomenon. According to Internet World Stats, as of December 2017, more than 4.1 billion people were online, up from almost 3.3 million people in 2015. This represents more than 26% growth in just two years and a 1,052% increase since 2000.

Figure 2.4 shows how many people use the internet worldwide.

* Asia has the highest number of internet users, at more than two billion (almost half of the world's total users).
* North America, including the US and Canada, comes in fifth with 345,660,847 users.

FIGURE 2.4 **Worldwide Internet Use**

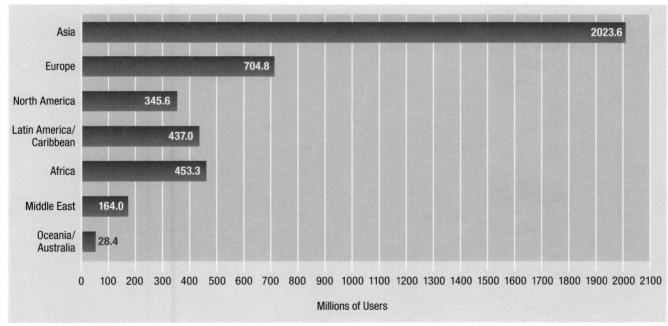

Source: http://InternetWorldStats.com, December 2017.

However, these statistics can be misleading. In fact, relative to the size of our population, North America has the greatest penetration of internet usage: 95% of us are online, compared to only 48.1% of the population of Asia.

To appreciate how the online world works, it's useful to understand how we all connect to this vast community with a combination of hardware and software technology.

Connecting to the Internet

Most of today's computers come ready to connect to the internet through the use of wireless technology. After you establish an account with an **internet service provider (ISP)**, which is a company that lets you use its servers and software to connect to the internet for a fee, you will need hardware such as a DSL, cable, or satellite modem to connect (see Figure 2.5). If you want to connect wirelessly, you will need additional hardware such as a router.

FIGURE 2.5 **Connecting to the Internet**
Depending on the type of internet connection available, you will need different types of equipment to connect to the internet.

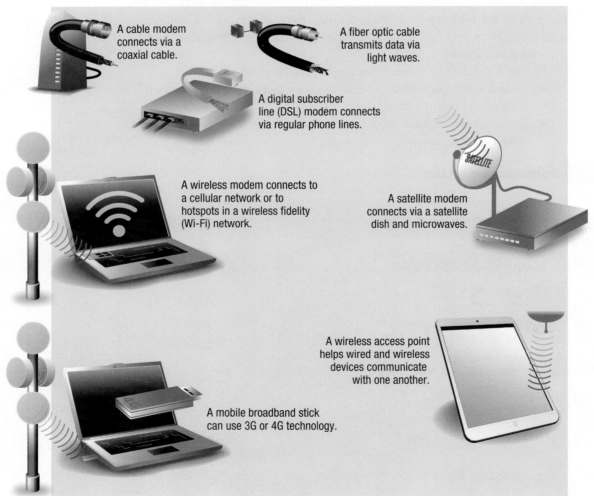

A cable modem connects via a coaxial cable.

A fiber optic cable transmits data via light waves.

A digital subscriber line (DSL) modem connects via regular phone lines.

A wireless modem connects to a cellular network or to hotspots in a wireless fidelity (Wi-Fi) network.

A satellite modem connects via a satellite dish and microwaves.

A wireless access point helps wired and wireless devices communicate with one another.

A mobile broadband stick can use 3G or 4G technology.

You might connect in a number of ways:

- You can get access via a cable modem that connects your computer via a coaxial cable (the same type of cable that carries a cable TV signal), by using a high speed digital subscriber line (DSL) modem that connects your computer via regular phone lines or by using fiber optic cable, which transmits your data via light waves. You can share any of these connections over a wireless router.
- Use the connection from your mobile phone to connect to the internet. This is called **tethering**.
- Use a wireless modem (most newer computers offer this option) to go online by way of a cellular network.
- Tap into a wireless fidelity (Wi-Fi) network, which uses radio signals for a wireless connection from hotspots located in places such as hotels, airports, and internet cafés. These may be free or may require that you pay a fee to connect. Some ISPs now offer a wireless cloud over their service areas, making wireless access available in open places like your hometown's parks and recreation areas.
- **Multihomed devices** allow you to connect to the internet in multiple ways. For example, your smartphone can connect using your cell plan or a Wi-Fi connection. The best method for connection is determined based on those currently available, and if the existing connection becomes unavailable (such as if you leave the area of a Wi-Fi network), the device will automatically switch to another connection, if possible.
- Mobile broadband sticks are essentially portable modems about the size of a USB stick. They use 3G or 4G technology to connect via a provider such as T-Mobile or Verizon for a monthly fee. 3G is the third generation of cellular technology. 4G is fourth generation technology that provides increased speed and bandwidth. Both 3G and 4G technology are also used in smartphones.

A smartphone is a multihomed device, which allows you to access the internet using a cell plan or a Wi-Fi connection.

You may still occasionally come across an old-fashioned modem that is used for a low-speed connection method called *dial-up*. With a dial-up internet connection, you plug a phone line into your computer and dial a local access number provided by your ISP to go online. The connection is usually slow and you can't use your phone for calls while you're connected.

Paying for the Privilege: Internet Service Providers

As mentioned earlier, an internet service provider (ISP) lets you use their technology (servers and software) to connect to the internet for a fee. Your ISP might be your phone or cable company. There are also national ISPs such as Sprint and local and regional ISP companies that use technologies such as DSL and fiber optic access provided by local utility companies.

Depending on your connection method, your ISP may provide you with a modem and/or router and instructions for using it to make your connection. ISP accounts typically include a variety of services, such as an email account, news and other information services, and security services such as a firewall, virus scans, and spam management for your monthly subscription fee. Email and information services may be identified and displayed on your ISP start page, which is often personalized with your name and a weather report for your area. You can personalize your start page further—for example, by choosing which news services send news to the page.

How Browsers View What's Online

You use browser software such as Microsoft Edge, Firefox, Safari, Google Chrome, and Opera (Figure 2.6) to view online content. Essentially, a **browser** renders web pages (which are typically created using a language such as HTML) into text, graphics, and multimedia. As of July 2017, Chrome was the most popular browser with 60% of market share.

Browsers allow you to move from page to page on the web while retaining a list of your favorite sites and a record of your browsing history. When you find a file online that you want to **download** to your computer, you can do that using tools in your browser. You can also use your browser to **upload** a file to store in the cloud. With built-in filters and security features, browsers have become an ally in protecting your privacy online.

Mobile devices use web browsers designed to display web pages on smaller screens. For example, Opera Mobile allows a smartphone running the Windows, iOS, or Android operating systems to browse the web on the go. (Chapter 4 provides more information about mobile phone operating systems.) As many people now surf the web from

Smartphones allow you to browse the web with mobile versions of your favorite browsers, such as Safari, Chrome, and Opera.

FIGURE 2.6 **Google Chrome**
Google Chrome is the most popular browser today.

Back button

Address bar

Our Digital World

their mobile devices, more companies are providing mobile apps and responsive websites. Responsive websites contain all or most of the same content, but they are specially formatted to be viewed and interacted with more easily on small screens.

Course Content
Take the Next Step Activity
Ethics and Technology Blog

2.4 Navigating and Searching the Web

With so many web pages out there created by lots of different people, the web ought to be pretty chaotic. However, there are underlying systems as to how pages are organized on the web and how they are delivered to your computer. This involves unique addresses used to access each web page, a unique address for each computer, and browser features for locating and retrieving online content.

IPs and URLs: What's in an Address?

An **Internet Protocol (IP) address** is a series of numbers that uniquely identifies a location on the internet. An IP address consists of four groups of numbers separated by periods; for example: 225.73.110.102. A nonprofit organization called *ICANN* keeps track of IP addresses around the world.

Due to the explosion of internet users and websites, the number of available unique four-group addresses (under IPv4) has been consumed. A new standard known as IPv6 uses IP addresses with eight groups of hexadecimal characters separated by colons and should provide enough addresses for the foreseeable future.

> The goal has always been the same. The progression is from data to useful information to knowledge that answers questions people have or helps them do things. Knowledge is the quest.
>
> —Amit Singhal, senior vice president at Google

Because numbers would be difficult to remember for retrieving pages, we use a text-based address referred to as a **uniform resource locator (URL)** to go to a website (Figure 2.7). A URL, also called a **web address**, has several parts separated by a colon (:), slashes (/), and dots (.). The first part of a URL is called a *protocol* and identifies a certain way for interpreting computer information in the transmission process. **Hypertext Transfer Protocol (HTTP)** allows you to view a site, transferring information back from the server. This is the typical protocol used when you visit a website. **Hypertext Transfer Protocol Secure (HTTPS)** encrypts communications using a secured connection. This helps ensure that your communication is seen only by the intended recipient. **File Transfer Protocol (FTP)** enables the uploading and downloading of files. Some sites use a secondary identifier for the type of site being contacted, such as *www* for a World Wide Web site, but this is often optional.

The next part of the URL is the **domain name**, which identifies the group of servers (the domain) to which the site belongs and the particular company or organization name (such as *ParadigmEducation* in the Figure 2.7 example).

A suffix, such as *.com* or *.edu*, further identifies the domain. For example, the *.com* in the example shown in Figure 2.7 is a **top-level domain (TLD)**. These TLDs

FIGURE 2.7 Parts of a URL

A domain name is an easy-to-remember way to get to a specific IP address.

with three or more characters are also called **generic top-level domains (gTLDs)**, in contrast to two-character country code TLDs such as *.jp* and *.de*. To accommodate growth, ICANN is delegating new gTLDs on a rolling basis and expects to eventually move the domain name system to over 1,300 new names. Table 2.2 provides a rundown of some common gTLDs; note that some four-character gTLDs have begun to appear.

Browsing Web Pages

You may already be quite comfortable with browsing the internet, but you may not have pondered how browsers move around the web and retrieve data.

Any element of a web page (text, graphic, audio, or video) can be linked to another object or page using a hyperlink. A **hyperlink** points to a destination within a web document or on another website. Text that is linked is called **hypertext**.

A website is a series of related web pages that are linked together. You get to a website by entering the URL, such as https://amazon.com, into your browser. Every website

TABLE 2.2 Common Generic Top-Level Domain Suffixes Used in URLs

Suffix	Type of Organization	Example
.biz	business site	Billboard: https://www.billboard.biz
.com	company or commercial institution	Intel: https://www.intel.com
.edu	educational institution	Harvard University: https://www.harvard.edu
.gov	government site	Internal Revenue Service: https://www.irs.gov
.info	unrestricted	Patient: https://www.patient.info
.net	administrative site for ISPs	Verizon: https://www.verizon.net
.org	nonprofit or private organization	Red Cross: http://www.redcross.org

has a starting page, called the **home page**, which is the first page displayed when you visit the site's TLD. For example, the home page of the US Copyright Office is shown in Figure 2.8. You can also enter a URL to jump to a specific page within a site, such as the page showing copyright fees at https://www. copyright.gov/about/fees.html, or search within the site to find specific content.

Searching for Content Online

A **search engine**, such as Google, Ask, or Yahoo, catalogs and indexes web pages. A type of search engine called a **search directory** can also catalog pages by topic, such as finance, health, news, shopping, and so on.

Search engines may seem to be free services, but they are typically financed by selling advertising. Some also make money by selling information about your online activities and interests to advertisers.

Some current search engines, including Microsoft Bing and Google, not only search for content but make choices among content to deliver more targeted results. Such search engines will respond to a search for a list of female tennis stars from 1900 to present day by assembling and displaying the results in a table.

Table 2.3 shows some common search tools with their URLs.

TABLE 2.3	Common Search Tools
Search Tool	**URL**
Ask	ask.com
Bing	bing.com
Dogpile	dogpile.com
DuckDuckGo	duckduckgo.com
Google	google.com
Yahoo!	yahoo.com

Searches That Succeed So how do search engines work? You search for information by going to the search engine's website and typing in your search text, which is comprised of one or more **keywords** or keyword phrases. You can search within a specific website by entering your search text into the search text box, which you'll find on most websites.

FIGURE 2.8 **Parts of a Web Page**

Web pages are designed with tools to help you access their content.

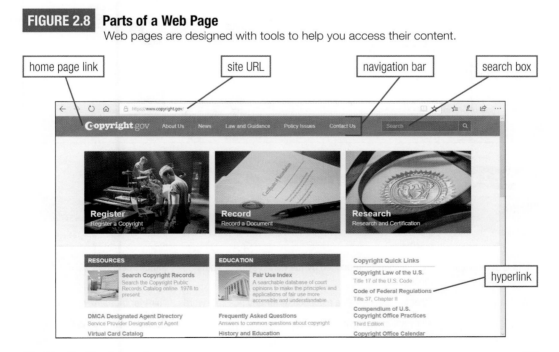

For example, to find information about the international space station, you might type *international space station* into the search engine's search text box and press Enter. You can then narrow your search by specifying that you want to view links to certain types of results such as images, maps, or videos, as on Google's site, shown in Figure 2.9.

You can get more targeted search results by honing your searching technique. Effective searching is a skill that you gain through practice. For example, typing *space station* into a search engine could easily return more than 80 million results. If what you really need to find out is the cost to build the station, consider a more targeted keyword phrase like *space station cost*. Search engines provide advanced search options that you can use to include or exclude certain results. For example, you can exclude pages with certain domain suffixes (such as *.com* and *.net*) to limit your search results to education and government sites. Table 2.4 explains how to narrow your search by entering keywords in various ways. Figure 2.10 shows Google's advanced search settings, which include options for search terms, search choices, language, region, SafeSearch, and last update.

A **metasearch engine**, such as dogpile.com, uses several search engines to search keywords across several websites at the same time. This helps optimize the search by providing the top results from the

FIGURE 2.9 The Google Search Page

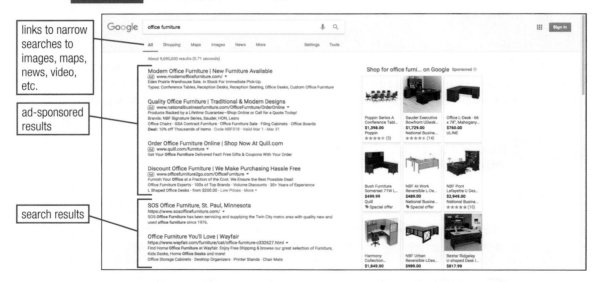

TABLE 2.4 Advanced Search Parameters

Item	What It Does	Example
Quotes ("")	Instruction to use exact word or words in the exact order given	"Pearl Harbor"
Minus symbol (-)	Excludes words preceded by the minus symbol from the search	jaguar -car
Wildcard (*)	Treat the asterisk as a placeholder for any possible word	*bird for blue*bird*, red*bird*, etc.
Or	Allow either one word or the other	Economy 2006 or 2007

FIGURE 2.10 Google's Advanced Search

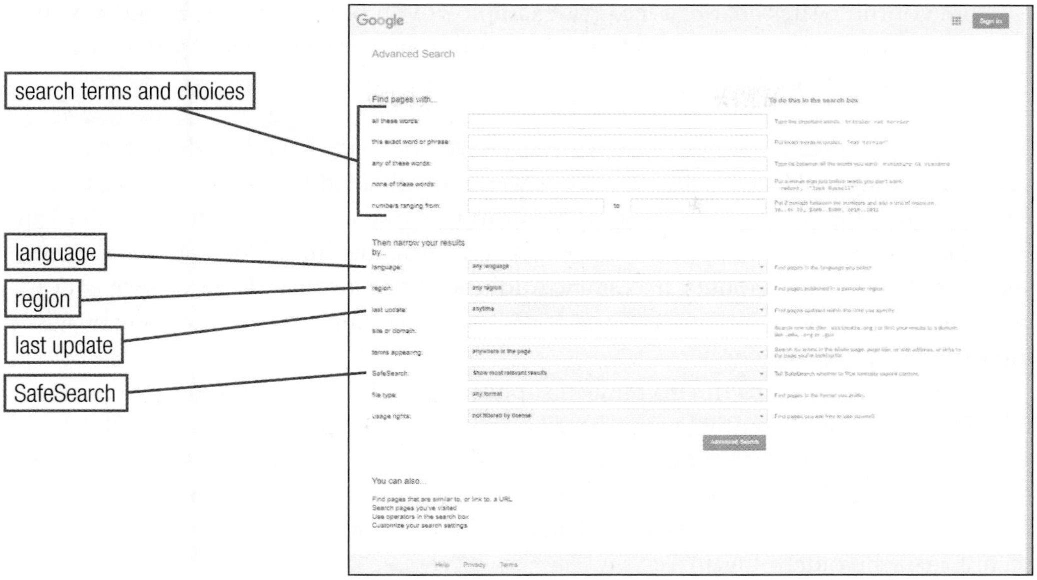

Labels on figure:
- search terms and choices
- language
- region
- last update
- SafeSearch

best search engines. For example, imagine you need to write a report on cocoa production. Instead of typing *cocoa production process* into several search engines individually, you can type the search text into one metasearch engine that initiates searches on several engines at the same time.

Specialized Search Sites Many websites are essentially specialized search services. Some simply provide you with information while others use your search results to sell you merchandise or services. Here are some examples:

- Mapquest and Google Maps let you look for locational information including maps and driving directions.
- Expedia is a travel site that searches a multitude of resources for rates on airfares, hotels, and rental cars to help you plan a trip. Similarly, KAYAK searches hundreds of travel websites so you can compare results and find the best deal.
- BizRate is a site you can use to compare and shop by gathering results for prices at dozens of retailers.

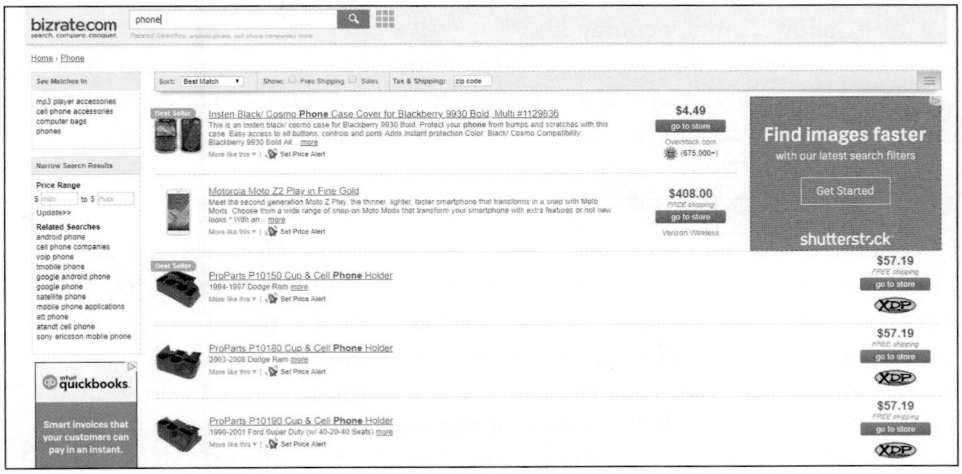

BizRate is a metasearch engine that allows you to compare information from several retailers to find the best price available for a product.

- Google Video indexes videos that have been posted online from around the world.

Some of these sites charge a fee. For example, if you book a trip on Expedia, you pay for the trip plus a small fee to Expedia for providing your itinerary. Other sites, such as Mapquest, are free.

Adding Capabilities to a Browser You may have heard the term **plug-in** or **player**. Historically, these apps were downloaded to your computer by a web page to support activities such as **streaming media** and viewing files. Much of the functionality provided by plug-ins is now natively available in HTML5 and CSS3, so specialized apps are no longer necessary. Additionally, security and speed were issues with plug-ins. As of 2017, browsers no longer support plug-ins, making your browsing experience more secure.

The one exception to the no plug-in support is the Adobe Flash plug-in. Support for Adobe Flash is currently being phased out and will be totally blocked by 2020.

Browsers also use extensions for certain types of specialized activity such as blocking advertisements and taking screen captures. An **extension** is downloaded once and extends the capabilities of a browser. The extension is run by the browser and is more secure than the older plug-ins.

Course Content
Take the Next Step Activity

2.5 The Vast Sea of Online Content

Though it's difficult to calculate exactly how many websites and web pages exist today, the Netcraft Secure Server Survey, reported over 1.7 billion hostnames in December 2017, up from 0.75 million just five years earlier. With that kind of constant activity, it's logical to conclude that not all online content is of the same quality or accuracy. In addition, some of that content is free for the taking, while other content is protected by the legal right of ownership called **copyright**. It's important that you learn how to evaluate the quality of content, learn to respect laws that govern your use of that content, and understand when free exchange of content is allowed and encouraged.

Some information posted online occurs in social transactions such as on blogs and social networking sites. Chapter 7 takes a look at the social web and the kinds of content and activity you'll find there.

Evaluating Web Content

The web contains a wealth of accurate and useful information. However, not all information found online is true, just as newspaper stories can contain inaccuracies. As in the offline world, you have to consider the source of online content. If you trust technology information from *Wired* magazine in print, you can have a similar level of trust in its online content. If you don't know a source at all, you may have to do some digging to discover if it is reputable by looking at the source's credentials (what individuals or organizations are involved in the venture?), methods (is the information based on surveys and experiment, or personal opinion?), and reputation (what do other online users say in reviews of the site or the company's products?).

Because anyone can publish to the web, to gauge the accuracy of what you read, you have to verify the three Ws (or WWW) of online content (Figure 2.11):

- *Who* is the author or publisher? Is the source credible?
- *What* is the message? Is the information verifiable? Is there a possibility of bias? Always try to crosscheck the information with other sources. Look for sponsors of a site to determine if the content could have a bias.
- *When* was it published? Is the information current? If no date is published, is it possible to figure out how current the information is from the text? Online information may remain available even after it becomes outdated. Always look for the most current information on any topic.

FIGURE 2.11 **The Three Ws of Online Content**
Use *Who*, *What*, and *When* information to gauge the accuracy of content found on the web.

Some academics prefer more authoritative sources than the collaborative Wikipedia. Still, many people find the shared knowledge it provides innovative and valuable.

The Wikipedia unified mark is a trademark of the Wikimedia Foundation and is used with the permission of the Wikimedia Foundation. We are not endorsed by or affiliated with the Wikimedia Foundation.

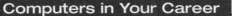
A wide variety of careers rely heavily on using the internet for research. Librarians, government policy analysts, economists, insurance risk analysts, and others make use of online encyclopedias, survey results, and online professional journals on a regular basis. Human resource workers search online for the latest compensation models for jobs in different areas of the country. Purchasing agents search for vendors and the best prices online. What career are you considering and how could online research help you succeed in that field?

Intellectual Property

Some works found online are placed there to be shared and passed on. Other content falls into the category of intellectual property, much of which is copyrighted. According to the World Intellectual Property Organization, **intellectual property (IP)** refers to "the creations of the mind, such as inventions; literary and artistic works; designs; and symbols, names, and images used in commerce." It is illegal to copy or distribute intellectual property without appropriate permission.

> You can't say that you work in research and not worry about IP because as soon as you do any research you are doing intellectual property – it goes with the research and you have a responsibility to look after it.
>
> —Dr. Philip Graham, Director of the Association for University and Industry Links

The internet has brought the issue of illegal use of intellectual property front and center. Because it's so simple to copy and paste content online, many people who would never dream of stealing a CD from a music store or a book from a bookstore download music illegally or plagiarize by using text or images from a website and representing that content as their own work.

In the early 2000s, peer-to-peer (P2P) file sharing was common. P2P permits users to download files from one another rather than through an online server. In addition to copyright violations, risks associated with P2P file sharing included identity theft and downloading of malware. A Supreme Court decision in early 2005 led to the closure of many P2P services and an end to the type of file sharing that was common at that time. P2P has largely been replaced by limited use free versions of commercial software, subscription software (Office 365), and media streaming services such as Spotify and Netflix.

Some people feel that copyright law in the digital age has gone too far. There's a strong sense among many internet activists that laws such as the Digital Millennium Copyright Act distort the balance between fair use (the right to reuse content that is available to all) and intellectual property rights and therefore are a threat to creativity and technological innovation.

The Invisible Web (aka the Deep Web)

The content you typically find online is only the tip of the web-content iceberg. Pages visible in standard web searches are typically indexed using automated search tools called **spiders**, which build an index based on words found on web pages. However, there are huge "hidden" collections of information that are collectively known as the **invisible web** or **deep web**. A typical search engine won't return links to these databases or documents when you enter a search keyword or phrase. Many search engines use **crawlers** that follow a long trail of hyperlinks to find data. However, many databases are set up to only respond to typed queries, so the crawlers strike out. Also, to get some databases on the invisible web, libraries and companies have to pay for access to them.

In the future, as search engines get more sophisticated, you will probably be able to more easily find this content. You can try now by entering the word *database* after your keyword(s) in services such as Google. Doing this may help you locate some hidden content. You can also try to locate hidden content in directories such as the Librarians' Internet Index, free and paid-for databases such as LexisNexis for legal research, and some specialized search engines such as Science.gov. Science.gov searches over 60 databases and over 2,200 science-oriented websites to provide users with access to more than 200 million pages of US Federal Government–provided scientific information, including research and development results. If you're willing to pay for help accessing the invisible web, companies such as BrightPlanet specialize in harvesting information.

Data integration is an important factor in obtaining more complete search results. With data integration, a search engine cross-references the records of one database with another, so your answer is much more complete. For example, you could cross-reference a search on the Supreme Court with a legal database.

> **Course Content**
> Take the Next Step Activities

2.6 E-Commerce

Electronic commerce, or **e-commerce**, involves using the internet to transact business. When you're buying downloadable music or software, shopping for shoes, or paying to access your credit report, you're involved in e-commerce.

There are three main kinds of e-commerce that describe how money flows in an online business. Money can flow from business-to-consumer (B2C), business-to-business (B2B), or consumer-to-consumer (C2C). Sometimes more than one of these models occurs on a single site—for example, when a consumer on eBay buys a product from another consumer (C2C) and eBay also makes money from advertisers (B2B).

B2C E-Commerce

Business-to-consumer (B2C) e-commerce is probably the kind of online commerce with which you are most familiar. It involves companies, such as Amazon and Zappos, that sell products and services to individual consumers. This is the model that most resembles shopping for books or shoes at brick-and-mortar stores. The steps in the B2C online shopping process are shown in Figure 2.12.

B2B E-Commerce

Business-to-business (B2B) e-commerce involves businesses selling to businesses. In some cases, a business provides supplies or services to another business, such as a plumbing supply site that caters to building contractors.

FIGURE 2.12 **The B2C Online Shopping Process**

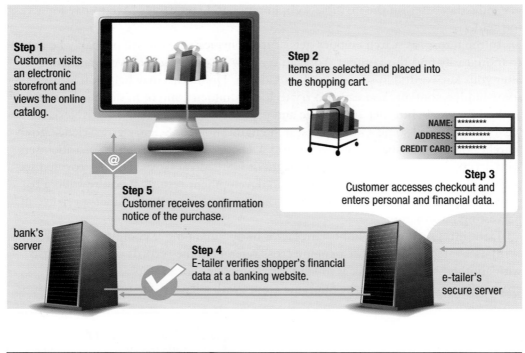

Step 1
Customer visits an electronic storefront and views the online catalog.

Step 2
Items are selected and placed into the shopping cart.

NAME: ********
ADDRESS: *********
CREDIT CARD: ********

Step 3
Customer accesses checkout and enters personal and financial data.

Step 5
Customer receives confirmation notice of the purchase.

bank's server

Step 4
E-tailer verifies shopper's financial data at a banking website.

e-tailer's secure server

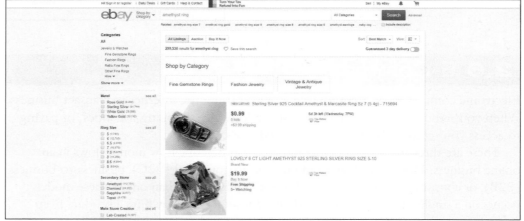

eBay is a popular auction site with B2B, B2C, and C2C characteristics because transactions can occur between two consumers, two businesses, or a business and a consumer.

In another B2B model, businesses provide a service to consumers but do not charge those consumers directly. Instead, their business model involves making money from selling ad space or information about their customers to advertisers. Given that e-commerce models are defined by how money flows, Facebook is an example of this second kind of B2B site because it gets no money from its members, only from advertisers (other businesses).

C2C E-Commerce

Consumer-to-consumer (C2C) e-commerce activity occurs on sites such as craigslist and eBay, where consumers buy and sell items from each other over the internet. Though the host site provides the infrastructure, the money flows from one consumer to another. What e-commerce model do you think supports the companies that host C2C sites? If you guessed B2B (they get their money from advertisers), you'd be right!

E-Commerce and Consumer Safety

In many cases, buying and selling items online is safer than doing so offline. That's because, rather than handing your credit card to a clerk in a store, you are performing a transaction over a secure connection, providing payment information to a system rather than an individual. Of course, every system has its problems, and online stores, banks, and investment sites are hacked into now and then. Still, if you use care in choosing trusted shopping sites, pay by a third-party payment service such as PayPal or by credit card (these purchases are protected from theft, while a check or debit card purchase is not), and make sure that while performing a transaction the URL prefix reads *https* (indicating a secure connection), you can be fairly confident that you'll have a safe shopping experience.

🖥 Computers in Your Career

Many people have become online entrepreneurs by starting their own eBay store or creating a website about a special interest such as baking bread. Others become affiliates or associates; they get a small fee for sending online browsers to a partner site. Some post videos on YouTube earning money from embedded ads. Whether you use e-commerce to supplement your regular job or as your main source of income, it can offer challenges and rewards.

☁ Course Content
Take the Next Step Activity

2.7 Connecting in Cyberspace

The internet offers a variety of ways to communicate that people are using to get work done or stay in touch with family and friends. We connect and collaborate in the workplace using multiple tools including devices such as our mobile phones and apps that run on corporate project websites. We use cloud storage and cloud apps to collaborate with others. Some companies provide a virtual private network (VPN) that permits individuals to access company data from off-site locations. This has helped to support work-at-home and flex schedules.

> 66 The internet is just a world passing around notes in a classroom. 99
>
> —Jon Stewart, comedian

The typical office space is no longer a paper-based environment with in-person meetings. It is now a digitally connected environment that supports shared access to content, live chats, and online meetings. Email, messaging, and conferencing have all helped to advance this new work model. In addition, social networking, which will be covered in Chapter 7, has also changed our world at work, in the classroom, and in our personal lives.

Email

Electronic mail (**email**) was one of the first services available on the internet, even before consumers started coming online and the web appeared on the scene. In fact, the first email was sent between two computers in 1971. Following

The computer used to send the first email in 1971 was huge by today's standards.

the advent of the commercially available internet, email soon became the standard method of communication for businesses and individuals.

Today, email is still extensively used in business even as the use of instant messaging, social networking, and chat is increasing. Email use is expected to grow by 4.4% per year through 2021. Email is favored for business communication by 86% of professionals. Today's average office worker receives 121 emails and sends out 40 emails a day. With today's mobile workforce, 66% of email is now read on mobile devices. In addition to appropriate, work-related email, almost 50% (49.7%) of email is spam.

You can use either **web-based email** or an **email client** such as Outlook to access your email. Both permit you to create and send messages, add attachments, receive and reply to messages, store contact information, and manage messages in folders. Most email services today allow you to manage multiple email accounts and sync email, contacts, and calendars across multiple devices.

An email client first converts an email into Multipurpose Internet Mail Extensions (or MIME) format. Years ago, email programs only supported the characters found on US keyboards. **MIME** is a standard that enables email to include other character sets and non-text attachments (such as media files, spreadsheets, PDF files, and so on). The email client next uses a set of rules, called the **Simple Mail Transfer Protocol (SMTP)** to send a MIME-formatted message to an **email server**, a computer system that sends and receives email. To retrieve an email, the email client uses either **POP3** (which stands for Post Office Protocol, Version 3) or **IMAP** (which stands for Internet Message Access Protocol).

You may have already observed that email services use a specific address format for sending messages, but you may not know what each piece of that address means. An **email address** includes a user name, the domain name for the email service, and the domain suffix, as in YourName@gmail.com. The user name and domain name are separated by the @ symbol.

Instant Messaging

Messaging between mobile phones and other portable devices is another growing form of communication. When you send a **text message**, you are exchanging a written message of 160 characters or less with another person using **short message service (SMS)** through a cell phone provider. Sending a text message is called **texting**. Text messages longer than 160 characters are sent in chunks and rebuilt by your cell phone carrier. The text message celebrated its 20th birthday in 2012. Approximately 2.2 trillion text messages were sent that year in the US, and 8.6 trillion messages were sent worldwide.

The growth of texting has slowed in recent years in favor of mobile instant messaging. An **instant message (IM)** is a message transferred between devices over a network. A **mobile instant message (MIM)** is a message transferred between mobile devices using Wi-Fi. Both instant messaging and mobile instant messaging require that the sender and receiver have compatible messaging app software. Messaging apps like iMessage, Skype, WhatsApp, Snapchat, and Google Hangouts and messaging services within social networks, such as Facebook Messenger, are taking over this form of communication. Figure 2.13 shows the usage of the most popular messaging apps in the US in November 2017. The use of mobile instant messaging is expected to increase from over 1.4 billion accounts in 2014 to over 3.8 billion accounts by the end of 2018. Mobile messaging permits users to make voice and video calls and to share media.

FIGURE 2.13 **Popular Mobile Messaging Apps**
According to a survey from November 2017, the three most popular mobile messaging apps in the US by monthly active users were Facebook Messenger, Snapchat, and WhatsApp.

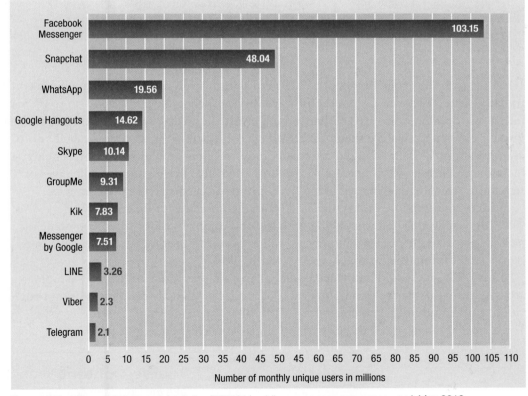

Source: https://www.statista.com/statistics/350461/mobile-messenger-app-usage-usa/, May 2018.

Messaging is increasingly used in business communications as it is more secure than email, provides the ability to break up a meeting into sub-groups (using individual chat rooms), and permits anyone to host an online meeting without subscribing to expensive software.

Audio and Video Conferencing

Audio and video conferencing are used more and more in the workplace for meetings and to support teamwork. Audio and video conferencing are often integrated into Learning Management Systems, which are used to electronically support education by providing an online classroom. Most conferencing systems can support both voice and video.

Video conferencing allows you to see the person or people, and even content, on the other end of a call.

Voice over Internet Protocol (VoIP) is a transmission technology that allows you to make voice calls over the internet using a service such as Skype. When you speak into a microphone, software and hardware convert the analog signals of your voice into digital signals that are then transferred over the internet in a process called **audio conferencing**. VoIP is also used for collaborative services such as **web conferencing**, a technology that allows companies to conduct interactive live meetings and deliver presentations over the internet with employee groups or clients across the country.

Video conferencing is very much like audio conferencing. It simply transfers video signals over the internet along with the audio signals. Since video conferencing permits all parties to see one another as well as hear one another, all participants must have a web cam on their devices.

Email, IM, and audio and video conferencing are only a few of the ways people are connecting using their computers and other devices. Today people are using text, video, audio, and images to collaborate with others in creative and exciting ways. You'll hear about many of these as you explore other chapters in this book.

> **Course Content**
> Take the Next Step Activities

Review and Assessment

Course Content
The online course includes additional
training and assessment resources.

Summing Up

2.1 The World Goes Online

Today more people are using the internet for a variety of activities over several kinds of devices, such as smartphones and tablets. You can **sync** devices to share data for calendars, contacts, and more.

There are many examples of cutting-edge uses of the internet in fields such as science, medicine, and even space travel. People are influenced in their purchasing decisions by the internet. **QR codes** are turning up everywhere for you to scan from your smartphone to get information about businesses and historical locations.

In the future, we will see even more uses of technology, more social collaboration, and advances in everything from retail to healthcare to communication from space, all because of the internet.

2.2 What Are the Internet and World Wide Web?

The **internet** is the physical infrastructure, made up of many smaller networks, that allows us to share resources and communicate with others around the world. It is made up of hardware such as servers, routers, switches, transmission lines, and towers that store and transmit vast amounts of data.

The **web** is a body of content that is available as web pages. The pages are stored on servers around the world. A **web page** may contain text, images, interactive animations, games, music, and more. Several web pages make up a single **website**.

The web continues to evolve, and the different phases of the web represent shifts in online usage and technologies. **Web 2.0** has become associated with interactive web services such as Wikipedia and Facebook. These services allow users to collaborate— share, exchange ideas, and add or edit content. Where Web 2.0 focused on *exchange* of data by individuals, **Web 3.0** (also called the **Semantic Web**) is evolving and will involve *integration* of data from a variety of sources in a meaningful way. Web 3.0 introduces the ever-present web and brings smarter applications that will evolve into an artificial intelligence that understands context rather than simply comparing keywords. It is also thought that Web 3.0 will help grow the **Internet of Things (IoT)**, enabling data transfer over a network without human involvement.

2.3 Joining the Digital World

You can connect to the internet through various methods, such as through a broadband cable, satellite, or DSL phone connection. An **internet service provider (ISP)** lets you use their hardware to connect to the internet for a monthly fee. To go online, you need an internet account, which generally involves connecting a piece of hardware such as a cable, DSL, or satellite modem, and often a wireless access point and/or router to share the connection. With newer computers, hardware and software for connecting to the internet may be built in, as with wireless technology. You can also use the connection from your mobile phone to connect to the internet via **tethering**.

A **browser** such as Google Chrome is what you use to view online content. Essentially, a browser renders web page content, which is typically created using a language such as hypertext markup language (HTML) into text, graphics, and multimedia. Browsers also allow you to navigate the web.

2.4 Navigating and Searching the Web

An **Internet Protocol (IP) address** is a series of numbers that uniquely identifies a location on the internet. Current IP addresses consist of four groups of numbers, IPv4, separated by periods. New addresses with eight groups of hexadecimal characters, IPv6, separated by colons, will ensure a pool of future addresses. Because numbers would be difficult to remember for retrieving pages, we use a text address referred to as a **uniform resource locator (URL)** to go to a website. The text-based addresses are cross-referenced to IP addresses.

To get to a website, enter a URL, such as https://amazon.com, in the address bar of the browser. Every website has a starting page, called the **home page**. Any element of a page (text, graphic, audio, or video) can be linked to another web page using a hyperlink. A **hyperlink** describes a destination within a web document and can be added to text or to an image such as a company logo. Text that is linked is called **hypertext**. Two protocols, **Hypertext Transfer Protocol (HTTP)** and **Hypertext Transfer Protocol Secure (HTTPS)** permit transfers between a web site and a server. Navigating among pages using a browser is called *browsing the web*.

A **search engine** allows you to search for information by going to the search engine's website and typing one or more **keywords** or keyword phrases.

Historically, some components on a web page required that you install a **plug-in** or **player** on your computer to view or hear content. Today, browsers no longer use plug-ins. Much of the functionality provided by plug-ins is now natively available in HTML5 and CSS3 or can be added as a browser **extension**. New web browsers and extensions have added security and speed.

2.5 The Vast Sea of Online Content

As in the offline world, you have to consider the source of content online. You should verify the three Ws (who, what, and when) of online content. Look at the credibility of who has written and published the information, the quality of the information, and how current it is to evaluate its quality and accuracy.

Some online content is free to exchange, share, and remix. **Intellectual property (IP)**, which includes inventions and literary and artistic works, cannot be copied or distributed without permission. Much of this material has the additional legal protection of a **copyright**. The internet can make it easy to copy and use other people's property, so intellectual property abuse is rife.

Data integration is a search feature that allows you to cross reference two web sites for search results. A **crawler** enables a search engine to follow a long trail of hyperlinks. Databases and other collections of information on the web that are hidden or not catalogued by most search engines are collectively known as the **invisible web** or **deep web**. To get to some databases on the invisible web, libraries and companies have to buy access to them.

2.6 E-Commerce

Electronic commerce, or **e-commerce**, involves using the internet to transact business. E-commerce sites are classified based on the flow of money: **business-to-consumer (B2C)**, **business-to-business (B2B)**, and **consumer-to-consumer (C2C)**.

2.7 Connecting in Cyberspace

You use an **email** program to create and send messages, add attachments, receive and reply to messages, store contact information, and manage messages in folders. **Multipurpose Internet Mail Extensions (MIME)** is a standard that enables email to include various character sets and non-text attachments. **Simple Mail Transfer Protocol (SMTP)** provides a set of rules for sending emails.

A **text message** is a brief message of up to 160 characters sent using **short message service (SMS)** through a cell phone provider. An **instant message (IM)** is transferred between devices over a network, while a **mobile instant message (MIM)** is transferred by a mobile device over Wi-Fi. Both instant messaging and mobile instant messaging require that the sender and receiver have compatible message app software.

Voice over Internet Protocol (VoIP) is a transmission technology that allows you to make voice calls over the internet using a service such as Skype.

Using the internet for phone meetings is called **audio conferencing**. **Video conferencing** allows you to see, as well as hear, the conference participants and content.

Web conferencing is a service that combines several technologies such as VoIP and webcams to allow users to view presentations and participate interactively while conducting live meetings over the internet.

Terms to Know

2.5 The Vast Sea of Online Content

2.6 E-Commerce

2.7 Connecting in Cyberspace

Projects

Check with your instructor for the preferred method to submit completed work.

Smartphones—Staying Well While Staying Connected

Project 2A TEAM

Play the interview with Dr. Eric Topol at https://ODW5.ParadigmEducation.com/iDoctor. Consider the ways he suggests that smartphones can be used to replace expensive medical treatments, and then respond to the following:

1. List three uses for smartphones that are demonstrated by Dr. Topol in the video.
2. What does Dr. Topol say about the future possibility of using a smartphone app to replace a visit to the doctor?
3. Do you believe your smartphone can be used to replace your physician? Defend your answer.

Individually or in teams, prepare for a debate. Can smartphones replace physicians? Your instructor will assign you to the pro or con position.

Project 2B

Texting while driving is a safety hazard. Countries around the world have acted to help keep roads safe by enacting laws that prohibit texting while driving. According to the FCC (US) article at https://ODW5.ParadigmEducation.com/TextDriving, distracted driving kills more than 8 people and injures 1,161 daily. Texting and making or receiving calls using a handheld cell phone while driving is illegal in many locations. Search online to learn about texting and driving laws in your location. If you live in the US, visit https://ODW5.ParadigmEducation.com/Map and find your state. Is it illegal to use a handheld cell phone while driving in your state? Visit https://ODW5.ParadigmEducation.com/Penalties and view the penalty for distracted driving in your state. What is the penalty for texting while driving in your state? Do you think this is an appropriate penalty? Prepare a three-paragraph letter to be sent to your state's governor that explains your position on smartphone use while driving and the related penalties in your state.

Net Neutrality—Should Everyone on the Internet Be Considered Equal?

Project 2C

You have been assigned to research the principle of net neutrality. To start your investigation of this controversial topic, go to https://ODW5.ParadigmEducation.com/NetNeutrality and watch a video to learn about the federal appeals court's 2014 ruling on net neutrality, and visit https://ODW5.ParadigmEducation.com/NetNeutralityRepeal for information about the 2018 repeal of the regulations. At the time of this writing, Congress was considering restoring the net neutrality rules. Conduct an internet search to learn more about this principle. When you have finished, prepare a summary that defines net neutrality, covers its main concepts, and explains the controversy surrounding this topic. Include any necessary text citations and a list of internet sources.

Project 2D `TEAM`

Discuss with your team the results of your individual research from Project 2C. Next, address the following questions with team members: Should corporations be able to treat internet users differently depending on the services to which the users have subscribed? Why or why not? Based on your discussion, arrive at a joint conclusion that either supports or opposes net neutrality. Prepare a summary explaining your team's position and the information that led your group to reach this conclusion. Have one person submit the document to your instructor with all team members' names included.

Project 2E `TEAM`

Using your summary from Project 2D, prepare an oral presentation and a brief slide-show to present your team's position and rationale on net neutrality. First, write a script for an oral presentation and assign each team member a speaking part. Then create a series of slides to accompany your presentation. The slides should highlight the main points of your position and provide effective visuals. Give your oral presentation to your classmates and provide an opportunity for student discussion.

Investigating Climate Change

Project 2F

According to NASA, climate change is "a long-term change in the Earth's climate, or of a region on Earth." Use your favorite search engine to find two articles about climate change: one article should clearly support the science of climate change, and the other article should present opposing views. Apply advanced search techniques to find articles published after January 1, 2013. When you have finished locating and reading the articles, write a one-paragraph summary of each article and include the URL below each summary.

Project 2G

Evaluating the credibility and accuracy of articles on the internet is an important part of the research process. Using the three Ws (or WWW) technique you learned in this chapter, evaluate the credibility and accuracy of the two articles that you read for Project 2F. To your existing document, add a table after each article that includes your three Ws analysis of the article's content.

Project 2H TEAM

Each team member should search the internet to find a web page that has inflammatory or provocative content regarding climate change. Team members should share their web pages with their group and choose one of the pages to analyze for this project. As a team, write a one-page document that summarizes the content of the web page and offers the reasons for your team's decision that the content is unreliable or not credible or that the content is reliable and valid. Include a second page that lists the URLs and brief summaries of the other articles that were found by team members.

Project 2I TEAM

Prepare a short slide presentation for the class that displays the web page and then presents your reasons for the decision summarized in Project 2H.

Is It Legal?

Project 2J

Research what is meant by the term *cybersquatting*. Is it legal? What is the Internet Corporation for Assigned Names and Numbers (ICANN) doing to try to prevent the practice?

Using what you learned, prepare a letter to ICANN advising them of a hypothetical dispute with a particular URL and submit it to your instructor.

Project 2K

Learn about the Creative Commons license. Summarize what you have learned in one or two paragraphs. Locate one photo, one music file, and one video that are covered under Creative Commons licensing. For each, explain any restrictions you must consider if you wish to share the content. Submit the report to your instructor.

Project 2L TEAM

The Office of the US Trade Representative publishes a "Priority Watch List" that lists names of countries in which a culture of piracy fails to protect intellectual property. Divide into teams and locate the watch list on the web to answer the question: Which countries seem to have the weakest IP protection?

Discuss within your team how a business such as a software company should react when their product is routinely copied illegally in other countries. Do developing countries have a right to take software to help their economies grow? Summarize your main points and submit your team's summary to your instructor. Be prepared to present your summary to the class.

Class Conversations

Topic 2A   Is web research enough?

You are taking a sociology course and have been asked to write an essay on various kinds of bullying in high schools. Assume you conduct all of your research online, making sure that you use credible sources. Another classmate conducts all of her research at the library using traditional reference materials such as sociology journals and books.

Discuss the advantages of your internet-based research approach and the advantages of your classmate's library-based work. Should the two papers be considered equal when graded? Do you think others will view the journal- and book-researched paper as being of stronger academic quality? Why or why not? Is your web-researched paper likely to have a different perspective? Why or why not?

Optional: Create a blog entry that states your response to the previous questions and provides your rationale.

Topic 2B   How is YouTube changing society?

The popularity of YouTube has made it possible for anyone to become a media publisher and can make an average person suddenly famous. It seems that every week another video is posted that becomes a popular topic among friends and family members. Some of these videos are taken of other people with or without their knowledge.

Consider the following scenario. You are enjoying an afternoon with friends at a football game. You jump up to cheer and accidentally trip, causing the person next to you to dump a carton of popcorn all over you. You are not injured, but are somewhat embarrassed by the popcorn on your clothes and in your hair. Your friend, who thought you looked funny covered in popcorn, filmed the accident using a cell phone. While you head to the bathroom to clean up, your friend posts the video on YouTube.

a. What is the likelihood you would know this video existed on YouTube?
b. Should the friend who filmed the accident be required to get your release before posting the video? Why or why not?
c. Does this scenario violate any guidelines provided by YouTube?
d. If you became aware of the video, what would you do?
e. Should someone be allowed to capture video of you without your knowledge? If you answer no, how could society police this action?

Topic 2C   What's after Web 3.0?

If the evolution of the web makes it possible for intelligent agents to locate information for you on the internet based on your preferences and even update and modify your documents when you read a related article, what will that mean for the role of the internet in learning? If student work is updated automatically, what will be the role of individual contribution and how much inaccuracy could be introduced? What role will humans play in the development of information on the web in the future?

Topic 2D Should your email be private in your workplace?

Case law has upheld the right of an employer to read email sent to/from an employee's corporate email account. Read the FindLaw article called "Email Privacy" at https://ODW5.ParadigmEducation.com/EmailPrivacy. Discuss the advantages and disadvantages to an employee and employer when a company monitors email at work. How you will manage this lack of privacy when you enter the workforce or in your current job?

Chapter 3

Computer Hardware and Peripherals
Your Digital Toolbox

What You'll Accomplish

When you finish this chapter, you'll be able to:

3.1 Identify the major types of digital devices.

3.2 Recognize the components that make up a digital device and give an example of each part of a digital device.

3.3 Identify and differentiate between input and output devices that are used with today's digital devices.

3.4 List the components to assess and factors to consider when purchasing a digital device.

Why Does It Matter ?

We live in a time when digital devices rule. Whether you always have to have the latest gadget or are hesitant to adopt new technology, you still must realize that these devices, from the tablet you use to browse the web and email to the Xbox One you use to play games and stream videos, are an important part of your life. Understanding the parts of the hardware that make up these devices, how you get information in and out of them, and how to make wise choices when purchasing new technology can help you cope and prosper in our digital age.

 The online course includes additional training and assessment resources.

The tangible part of computing starts with hardware. Components for input, processing, output, and storage allow you to put data in and get useful information out.

Course Content

Take a Survey

Technology in Your Future Video

3.1 A World of Digital Devices

As you read in Chapter 1, digital devices differ from analog devices in that they use symbolic representations of data in the form of code and can process words, numbers, images, and sounds. Today the number of devices that fit that definition has expanded far beyond your desktop computer, but these devices still have certain things in common. They all have some form of memory and, however basic, an operating system; they provide a way to input and store data and output information; and they have a source of power.

Some digital devices have input and/or output features built in, such as with a laptop computer or smartphone. Others use **peripheral devices** that physically or wirelessly connect with them. For example, your desktop computer keyboard is a peripheral device used to input data, and your printer is a peripheral device used to produce output.

Digital devices come in many forms and sizes.

3.2 The Parts That Make Up Your Computer

Compare a typical desktop and laptop computer and you'll find that, although they may be packaged differently, they each contain similar hardware that processes data, provides battery power, offers ports and wireless connections for connecting to peripheral devices and networks, and stores data.

The Motherboard

If you were to open up your computer, you'd see that the **motherboard** is the primary circuit board (a board containing electronic connections). It holds the central processing unit, basic input/output system, memory, and other components (see Figure 3.1). The motherboard is really just a container where the various working pieces of your computer slot into one compact package. Circuits on the motherboard connect the components contained on it.

The central processing unit (CPU), which is a microprocessor (also called by the shorter terms *processor* or *core*), sits on the motherboard. The CPU processes a user's requests, such as opening documents or formatting text. A CPU is a thin wafer or **chip** made up of a semiconducting material, such as silicon. It contains an integrated circuit made up of a combination of miniaturized components and drives the system's computing capabilities. (Note that the term *central processing unit* is sometimes, albeit incorrectly, used to refer to the case that contains the processor and everything else within this case.)

A **multicore processor** contains more than one processing unit or *core*. By splitting up the work of computing tasks, multiple cores make your computer work faster. Multicore processors include dual core, which contains two processors, and quad core,

FIGURE 3.1   The Operations of the Motherboard

The motherboard is a container for various processors and expansion slots for cards that add functionality such as sound, graphics, and memory.

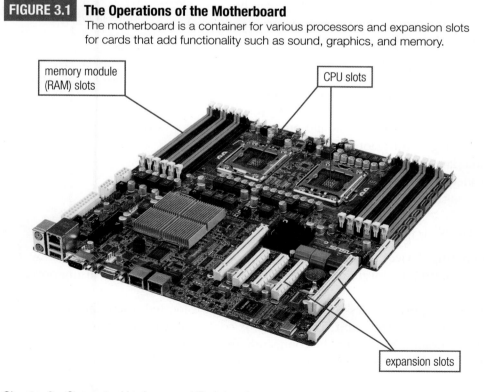

memory module (RAM) slots

CPU slots

expansion slots

which contains four processors. Computer chip manufacturers are starting to come out with six-core and eight-core processors. These cores may be integrated onto a single chip or be located on separate chips, in which case they may be referred to as multi-chip processors. How many cores you can fit on a chip essentially depends on how small the cores are. For example, quad cores were originally built on two chips, but as cores got smaller, those four cores were able to fit on a single chip.

All cores interact with the integrated memory controller on your computer, pulling data from memory to help them process tasks, and storing results of tasks back in memory. Cores may also have their own cache—a small high-speed memory storage area that holds the data that's most likely to be needed next—reducing the time it takes to access memory. (Cache memory was discussed in Chapter 1.)

Electronic connections between the processors and computer memory allow them all to interact. In the near future, processors may be able to connect using light, which will make your computer even faster.

Multicore processors are most efficient when used with software designed to take advantage of the processor's multitasking capabilities. Designing software in a way that allows tasks to be **parallelized** (run pieces on two or more processors at once) is important for better performance.

In addition to the processors, the motherboard also holds different types of memory. **Read-only memory (ROM)** is the permanent memory, or **nonvolatile memory**. ROM is hardwired into a chip. In a PC, the read-only memory stores the **BIOS** (basic input/output system). During the boot-up sequence, the BIOS checks devices such as your memory, monitor, keyboard, and disc drives to ensure they are working properly and then starts them up. It also directs the hard drive to boot up and load the operating system (OS) to memory.

Random access memory (RAM) chips are slotted into the motherboard and are used to store programs and data while the computer is in use. This memory is temporary, or volatile. Each memory location in RAM can be accessed in any order, which speeds up processing. (This differs from storage devices such as USB sticks or DVDs that store and retrieve data one file at a time.)

While some computers have built-in sound or graphics cards, others include them as **expansion cards**. These cards enable input and output of sound or images. Your laptop computer can also accommodate **PC cards** that slot into a built-in card reader to provide other kinds of functionality such as additional USB ports or wireless networking capabilities.

> "Researchers at IBM have created computer chips that behave more like actual brains when processing information. Systems built with these chips will be called 'cognitive computers.'… Cognitive computers are expected to learn through experiences, find correlations, create hypotheses and remember."
>
> —Rob Spiegel, TechNewsWorld

Power Supply

All computing devices require power to work, whether that power is accessed by plugging a cord into a wall outlet, operating off a charged battery, or using power from solar cells.

A **power supply** in a desktop or laptop computer is located where the power cord is inserted into the system unit. This metal box contains the connection for the power cord and a cooling fan to keep the connection from overheating. The power supply switches alternating current (AC) provided from your wall outlet to lower voltages in the form of direct current (DC). A laptop computer also contains a battery that is charged when you plug the power cord into your wall outlet.

Your operating system is capable of sending a signal to the power supply to instruct it to sleep or hibernate. This action puts the computer into a lower power mode or no power mode without losing your unsaved work. To take some computers out of sleep or hibernation mode, you can simply move the mouse or press any key. For most systems, however, you have to briefly press the power button.

Tablets and smartphones use USB chargers that can be connected to the device and then plugged into a wall outlet or a computer using an adapter. You can also purchase USB charger hubs, which are devices you plug into an outlet and then use to insert multiple USB connectors so that you can charge multiple devices at once. Charging from an outlet is much faster than charging from a computer.

A newer technology trend that is emerging is the charging mat. A **charging mat** or charging pad allows you to charge multiple devices simultaneously without connecting them to the mat with a cable. Charging mats typically use the same level of power that it takes to charge a device from an outlet. **Wireless charging** works with devices such as iPhone X to transfer an electromagnetic charge to a device from a charging station.

Charging mats are a way to charge more than one device at the same time.

Ports

A computer uses a **port** to connect with a peripheral device such as a monitor or printer or to connect to a network. A **physical port** connects a computer to another device, sending a signal via a cable (as with a USB port), infrared light (as with an infrared port), or a wireless transmitter (as with a wireless mouse or keyboard).

Some types of physical ports (commonly called just *ports*) in use today are serial, USB, FireWire, Thunderbolt, and infrared, as shown in Figure 3.2.

A **serial port** is a port, built into the computer, that is used to connect a peripheral device to the serial bus, typically by means of a plug with 9 pins. (A bus is essentially a subsystem of your computer that transfers data between the various components inside your computer.) Network routers in a business setting use serial ports for administration, although they are being replaced by web-based administration interfaces.

A **universal serial bus (USB) port** is a small rectangular slot that has become the most popular way to attach everything from wireless mouse and keyboard transmitters (the small devices that send wireless signals to devices) to USB flash drives for storing data. USB first came out in version 1.0, which supported a 12 megabits per second (Mbps) data rate (the measurement of the speed at which data can be transmitted). USB 3.0 became available in 2010, providing a transfer rate of up to 5 Gbps; and USB 3.1, introduced a few years later, provided even faster transfer, called *SuperSpeed*, with a data rate of 10 Gbps. USB-C, introduced in 2014, provides a small, reversible plug connector.

FIGURE 3.2 Ports in Your Computer

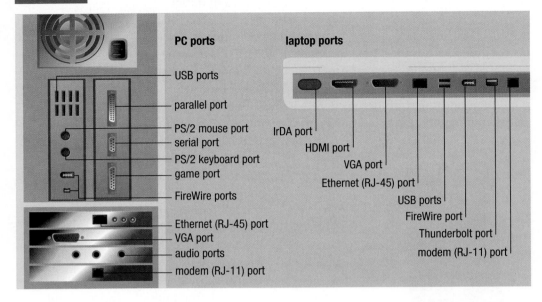

PC ports

- USB ports
- parallel port
- PS/2 mouse port
- serial port
- PS/2 keyboard port
- game port
- FireWire ports
- Ethernet (RJ-45) port
- VGA port
- audio ports
- modem (RJ-11) port

laptop ports

- IrDA port
- HDMI port
- VGA port
- Ethernet (RJ-45) port
- USB ports
- FireWire port
- Thunderbolt port
- modem (RJ-11) port

Usually a computer has two to four USB ports that you can use to attach peripheral devices. Should you need to plug in more peripherals than your computer has USB ports, you can add a hub that increases the number of available USB ports. Plug the USB hub into one of your computer's USB sockets and then plug USB devices into the ports on the hub. Most USB hubs offer four ports, but some offer more.

A **FireWire port** is based on the same serial bus architecture as a USB port. FireWire provides a high-speed serial interface for peripheral devices such as digital cameras, camcorders, or external hard disk drives. Generally used for devices that require high performance, the newest version of FireWire, called FireWire 800, is capable of transferring data up to 800 Mbps.

The **Thunderbolt port**, introduced on Apple's MacBook Pro in 2011, was developed by Intel and Apple to provide a peripheral connection standard that combines data, audio, video, and power within a single high-speed connection. The newest version of Thunderbolt, Thunderbolt 3, doubled bandwidth and cut power consumption in half. Thunderbolt's high-speed connection supports high-resolution displays and other devices that require a lot of bandwidth, such as high-end video cameras or data storage devices, and is generally expected to replace FireWire. Originally added to Apple devices, Thunderbolt is now being added to some PCs, making the port no longer strictly an Apple favorite.

An **Infrared Data Association (IrDA) port** allows you to transfer data from one device to another using infrared light waves. Today the ability to transmit wirelessly between devices using Bluetooth (a wireless communications standard) is making IrDA ports obsolete. You can add Bluetooth connectivity via a device that connects to the USB port.

MIDI is a communications protocol that allows computers and devices such as musical synthesizers and sound cards to control each other. Game ports on older computers allowed the MIDI device to be connected to a computer. A new version of MIDI that uses a high-definition protocol is being created. This HD MIDI protocol will accommodate more channels to carry data as well as create improved resolution of data values, allowing for clearer rendering of audio.

Wireless Connections

Most new computing devices have a built-in **wireless adapter** with the ability to detect and connect wirelessly to the internet if there is a Wi-Fi hotspot within range. Computers also have a standard wired network connection—variously called a NIC (pronounced "nick"), an RJ-45, or an Ethernet connector. Wired connections are generally faster than wireless connections.

Bluetooth wireless capability may also be built-in on your computer. If not, plug-in adapters are readily available. The technology can also be found in computer peripherals such as wireless mice, keyboards, and printers.

Storage

Because saving the work you've done on your computer is so important, computer manufacturers have devised several ways to store all kinds of data, from numbers and images to words and music. All storage media have methods for reading and writing data.

The various storage media on a computer are accessed using **drives**, which are identified on a Windows-based computer by a unique letter. For example, your computer's hard disk is typically identified as the C drive, while a DVD or USB drive might be labeled E, F, or G. Mac computers give each drive a name, such as Mac HD for the hard drive and DVD_VIDEO for a DVD drive into which you have inserted a video. If you are connected to a network, you may also be able to access shared network drives.

Hard Disks
The first place most of us save a copy of our work is on our **hard disk**, which is the disk drive that is built in to the computer. When you save a file to your Documents folder in Windows 10, for example, it is saved to your hard disk. The platters or disks in the drive rotate and one or more so-called "heads" read and write data to them. Because all hard disks eventually fail, the wise person uses other media storage or cloud storage to make backup copies.

Optical Drives
Your desktop computer is likely to have a built-in drive where you can place a **CD** or **DVD** to read content stored there or store data of your own. This type of drive is called an **optical drive**. To save on weight and size, many laptops today have no CD/DVD drive. Manufacturers assume you will download most of your software or content or use an external CD/DVD drive when necessary. Another type of optical drive that is built into some computers is a **Blu-ray disc** drive, mostly used for high-definition movies and games. Discs placed in optical drives are covered with tiny variations, or bumps, that can be read as data by a laser beam in the drive. Optical drives use these **optoelectronic sensors** to detect changes in light caused by the irregularities on the disc's surface. If you buy a movie or game on a disc rather than accessing the content online, it is likely to be in the Blu-ray format.

External Hard Drives
If you'd like additional storage, you might consider buying an **external hard drive**. External hard drives typically connect to your computer via a USB, FireWire, Thunderbolt, or Serial ATA cable, but select high-capacity models can connect to a network via a wireless connection. Some call these devices networked external hard drives, while others refer to them as **network attached storage (NAS)**.

You can connect to some external hard drives with USB or Thunderbolt cables.

External hard drives are useful for backing up your entire computer system. Some portable external hard drives don't even need a separate power source. Networked models can provide centralized storage for multiple teammates or family members or can serve as a "jukebox" for storing various forms of media such as music or videos.

Flash Drives Whether you call them **flash drives**, USB sticks, or thumb drives, these small devices are a convenient way to store your data and take it with you.

They come with capacities as big as 512 gigabytes (bigger than most hard drives just a few years ago), and some manufacturers have begun to release sticks whose capacity is measured in terabytes.

Flash drives use **flash memory** to record and erase stored data and to transfer data to and from your computer. Flash memory is also used in tablets, mobile phones, and digital cameras because it is much less expensive than other types of memory. Flash memory is nonvolatile, meaning that it retains information even in a powered-off state.

Flash drives are small, portable storage devices that are gaining larger capacities all the time.

Wireless Mobile Storage Devices Similar to a flash drive, a wireless mobile storage device allows you to store as much as 2 terabytes of data, wirelessly stream movies and music, and share content with your computer or other devices. This technology allows multiple devices to connect at the same time and can work over its own Wi-Fi network so no internet connection is required.

Solid-State Drives (SSDs) A **solid-state drive (SSD)** is essentially a flash-based replacement for an internal hard disk. These drives are lighter and more durable than traditional hard disks and have paved the way for smaller, more portable computers with longer battery. In 2007, SSDs began to show up in commercially available computers. One of the first was the XO Laptop (part of the One Laptop Per Child initiative in developing countries), and internal SSDs were also included in the new Apple MacBook Air and MacBook Pro released that year. Dell also began shipping an ultra-portable laptop with an SSD in 2007, and many other PC manufacturers have since followed suit, though the cost of SSDs has to come down before they will appear in most PC models. Intel's superfast Optane SSD includes a memory chip that could potentially be used for high-capacity storage. Optane SSDs have been measured as being 10 times faster than conventional SSDs.

Cloud Storage In addition to using physical storage, many users increasingly rely on cloud storage. Cloud storage refers to the storage of data on a web-connected server operated by a third party. You can access your files stored in the cloud from any location, using any computer with an internet connection. Cloud storage may be used for backing up information as well as sharing files with others. If you've ever lost or damaged your flash drive, you can appreciate how valuable it would be to have backup copies of your information that you could access via the web from your school, on a trip, or in your office. Depending on your needs, you can choose either a fee-based cloud storage service or free services such as Microsoft OneDrive. Some of the most popular cloud storage services today are Dropbox, Google Drive, Amazon S3, Box, Carbonite, iCloud, and iStorage.

☁ **Course Content**
Take the Next Step Activities

3.3 Input and Output Devices

Your computer is capable of processing and storing data, but if you can't get data into your computer and information out of it, this capability isn't of much use. That's where input and output devices come in. Examples of **input devices** are your keyboard to enter text or a microphone to record sound. **Output devices** produce information in one of several forms: printed text on a page, sound from a speaker, or an image on your monitor, for example.

What goes into your computer in the form of digital files may come out as printed text, a movie, an image, or sound.

A Wide Assortment of Input Devices

A computer **keyboard** contains keys that you press to activate electronic switches. These switches in turn tell the active software on your computer or in the cloud to insert numbers, letters, or characters in a document or form. Today's computer keyboards also allow you to combine some keystrokes to accommodate additional commands (the familiar Ctrl, Shift, or Alt combinations and function keys, for example). Some keyboards also include shortcut keys that allow you to manage media functions such as video or audio playback. Many users today also input data via onscreen keyboards, which require the user to switch among two or more keyboards to input letters, numbers, and symbols.

Key presses are interpreted by the software program that controls the keyboard, called a *device driver*. The operating system then provides the key press information to the currently active program, such as a word processor or an email client. Keyboards can plug into your computer, be built into a laptop, or use a wireless connection.

The option to use wireless input devices permits greater flexibility in positioning your keyboard and mouse.

Your **mouse**, also referred to as a *pointing device*, detects motion in relation to the surface you rest it on and provides an onscreen pointer representing that motion. You might think of a mouse as being like the lever that a crane operator moves to control the crane, which moves up or down in the air according to the movements of the lever. A mouse prototype was invented in 1964 at Stanford University. The first mouse was sold with the first Apple Macintosh in 1984.

A mouse can plug into your computer, be built into a laptop, or use a wireless connection. However it is connected to your computer, a mouse can function in several different ways. It can use:

- a mechanical device, such as a ball that rolls on a surface, to track motion.
- a light-emitting diode or **infrared (IR) technology** to sense motion, as with optical mice.
- an optoelectronic sensor that actually takes pictures of the surface.
- ultrasound technology to detect movement, as with 3-D mice.

Two other input devices are commonly used with today's digital devices: the touchscreen and the touchpad. A **touchscreen** is a visual display that permits the user to interact with a digital device by touching various areas on the screen, with either a finger or a **stylus**. A **digital pen**, such as the one that comes with Microsoft's Surface 4, is used to write or draw on a touchscreen. The touchscreen has evolved with the development of portable devices and smartphones. Touchscreens are used

in many different areas, from ATMs to airline kiosks. A **touchpad**, which senses finger movement, may be built into many computing devices. Both the touchscreen and the touchpad use the motion and position of a person's finger to locate a position on the computer screen, either directly (touchscreen) or indirectly (touchpad).

New input controls and interfaces are always appearing. For example, the Touch Bar on a MacBook provides controls for some input methods such as voice and formatting in a single bar at the top of the keyboard. The touch input panel on the Lenovo Yoga can be used as a virtual keyboard to draw or take notes with a stylus.

The **scanner** has a very descriptive name, as its function is to optically scan hard copy of text or images to convert them into electronic files. Scanners may sit on your desktop (called *flatbed models*), be built into an all-in-one printer, or be handheld. In the world of industrial design, 3-D scanners can scan all sides of objects and produce three-dimensional models of them.

Webcams are often used with online calling or meeting services so both callers can see each other as they chat.

A **webcam** is a video camera that can be built into your computer monitor or purchased separately and mounted onto your computer. Both webcams and digital cameras can become input devices when they interface with your computer to upload photos or videos in digital file formats.

Gaming devices such as Xbox One and Wii U provide controllers you can use to input moves in a game. The controllers often combine multiple buttons, a joystick, and even motion sensors. Virtual reality systems that are used to simulate situations for learning, such as for astronauts, offer **wired data gloves** that allow users to communicate with the system.

A **microphone** is an input device for getting sounds, from narrations to music, into your computer in the form of audio files. Microphones might be built into your computer, be plugged in via a cable, or be part of a headphone set.

Speech recognition software turns a user's spoken words into text. Leading voice recognition programs include Dragon Naturally Speaking and Dragon Dictate for Mac. The user typically completes a setup process to train the voice recognition software to understand his or her voice. The software then adjusts to the user's speech patterns, enabling that person to give commands and create documents such as spreadsheets and emails.

Virtual reality controllers allow a user to interact with a game environment.

A **mobile internet device (MID)** also uses a variety of input devices. MIDs may offer virtual keyboards, foldable keyboards, touchscreens, voice recognition, or styluses for providing input. There has been a great deal of debate regarding mobile devices being used in automobiles. Today, automobile and mobile device manufacturers are working to improve devices and software for hands-free use while driving. There are currently voice devices that enable a user to find local businesses and open, create, and respond to email and SMS messages totally hands-free.

Today, wearable devices are becoming more popular. Computers incorporated into watches, glasses, and clothing are always on. A user's skin, hands, speech, or even eye movement can be used to provide input. WPAN (wireless personal area network) and WBAN (wireless body area network) protocols are being developed so you can make your body's environment a moveable Wi-Fi network.

Assistive technology includes a variety of devices and methods that enable physically challenged computer users to control their computer and provide input. For example, sip-and-puff or wand and stick devices enable users to give computer input using their mouths. Other assistive devices include Braille embossers, screen readers, and speech synthesizers.

The Tongue Drive System is an example of assistive technology.

Retail and manufacturing employees often use bar code readers and RFID readers. **Bar code readers** optically scan a set of lines to identify a product. **RFID readers** scan an embedded tag that emits a radio frequency. Both are used to provide input to a computer system so software can track inventory and sales activity.

> ❝ Garbage in, garbage out. ❞
> —George Fuechsel, IBM technician, advising that bad input results in bad output

Getting the Most from Your Computer with Output Devices

Any device that displays, prints, or plays content stored in your computer is an output device, including your monitor, speakers, headphones, printer, and/or projector.

Monitors and Speakers You may not think of your **monitor** as an output device, but because it delivers information stored inside your computer in the form of images, it is. In the case of a mobile phone or gaming device, the screens are output devices. Monitor output is temporary; once you turn the computer off, there is no record of what was displayed.

Monitors may also include a **speaker** to add audio output. Laptop computers usually have an internal speaker, while desktop computers may use external speakers. Another way to get audio from your computing device is by plugging in a headset or by using a wireless **Bluetooth headset** or speaker.

Monitors come in a variety of display sizes, from a few inches on your smartphone to 10-inch screens on ultra-portable laptops to huge desktop screens that may be over 30 inches across. Display size is the measurement between two diagonally opposite corners. The latest USB 3.1 cables are enabling lightning-fast transfer of data such as high-definition 3-D video to your computer monitor or TV screen. **4K** screens are providing more vivid images for TVs, cameras, and monitors with almost twice the resolution of FullHD displays. **High dynamic range (HDR)** is making a similar difference for photos.

Modern computer monitors use one of these technologies:

- **Thin film transistor active matrix liquid crystal displays (TFT active matrix LCDs)** are the most prevalent type of monitor today. They use a thin film transistor (TFT) to display output from your computer.
- **Light-emitting diode (LED) displays** use light-emitting diodes. They conserve power and provide a truer picture than LCD models.

Some new display technologies beginning to appear on computer monitors and television sets include the following:

- **Surface-conduction electron-emitter displays (SEDs)** use nanoscopic electron emitters (extremely tiny wires, smaller than human hairs) to send electrons that illuminate a thin screen.
- **Organic light-emitting diodes (OLEDs)** project light through a thin electroluminescent (blue/red/green-emitting) film layer made of organic materials.
- 3-D or 3-D-ready technology uses a polarized lens to display images. Usually, the viewer must purchase special glasses and may have to install separate IR equipment for use with the computer in order to view the 3-D effect. This 3-D technology is being used for both computer monitors and televisions.

Bluetooth headsets let you talk on a mobile phone handsfree.

Printers and Faxes A **printer** is the main way in which you can get hard copy (print on paper) output from your computer. The foundation for today's printers was a dry printing process called *electrophotography*, also known as *Xerox* (hence the company of the same name). When the use of a laser beam was added, the business world saw the introduction of the laser printer in 1971 and inkjets (printers that spray jets of ink onto paper) in 1976. Hewlett-Packard threw its hat into the printer ring in 1984 with the first laser printer for sale to the general public.

LED displays are often used in television sets. Most computers can connect to a TV display to generate output.

A **photo printer** (which prints high-quality photos directly from a camera storage card) and a **thermal printer** (which heats coated paper to produce an image, like the kind you've seen printing receipts in retail stores) can also be used to create printed output. You may also encounter high-end commercial printers like the ones you see at your local copy store, and **plotters**, which are used to print large blueprints and other technical drawings.

You can use fax programs to send content from your computer that comes out as printed copy at the receiving end. If you use a **fax machine** to scan content and then send it, the machine converts the scanned content into an electronic file and then sends it to the recipient's fax machine, which prints it as a hard copy on the other end. In this case, the sending fax machine is an input device and the receiving fax machine is an output device.

3-D printing is a method of using printers to create 3-D objects from digital models. A 3-D printer prints layers on top of one another to build a precise, physical object, as opposed to traditional manufacturing, in which material is often removed to carve out an object. 3-D printing is being used to manufacture customized equipment, such as automotive parts, and patient-specific devices, such as orthopedic implants. The 3-D printing industry is expected to total $13.2 billion globally and will experience an annual growth rate of 22.3% with revenues reaching $28.9 billion in 2020.

Devices That Project Computer Content There are several different devices that can be used to project computer content as part of a presentation, either in person or across the web. An **LCD projector** projects light through silicone panels colored red, green, and blue. The light passing through these panels displays an image on a surface such as a screen or wall. By blocking or allowing light to pass through pixels on the panels, these projectors can output a huge range of colors.

If you give presentations, you will encounter LCD projectors. These output devices are popular for displaying PowerPoint slides and other types of content.

Used with a projector, a **document camera** can be used to display text, slides, a 3-D object, or any other printed material on a screen.

An **interactive whiteboard (IWB)** is a display device that receives input from a computer keyboard, special pen, finger, tablet, or other device. The IWB may connect directly to a computer to show the computer desktop, or you might need a projector to show the desktop on the IWB. IWBs help improve communication in settings such as the classroom, the corporate world, sports coaching, and broadcasting. Information displayed on an IWB may be saved as a document and shared or printed later. Popular IWB brands include SMART Board, Promethean, mimio, eBeam, and PolyVision.

Virtual Reality and Augmented Reality Displays

A **virtual reality system** connects you to a computer-simulated world. Most provide visuals to the user and some provide sound as well. The user wears a head-mounted display and headphones if sound is being provided. In the more sophisticated systems, gloves with wiring allow the user to control actions with his or her hands. Virtual reality is a connection between user and computer that allows both input and output. This technology is used to train pilots, astronauts, doctors, and others by having them deal with simulated situations.

Like virtual reality, **augmented reality** takes advantage of computer output. Augmented reality involves layering audio, video, and other output to improve our real world experiences.

An interactive whiteboard display device.

Virtual reality equipment like the Sony Morpheus headset for PlayStation is helping people in various industries to simulate and prepare for situations they might encounter in their work.

Course Content
Take the Next Step Activity
Ethics and Technology Blog

3.4 Purchasing a Computer

When faced with buying a new computer, many people get overwhelmed trying to figure out all the features and tech terms. The challenge is compounded because the specifications for what makes up the latest and greatest computing device change frequently as manufacturers try to outdo each other. However, there are certain questions you can answer to make an informed and intelligent computer purchase.

What Are Your Computing Needs?

Not everybody needs the newest, fastest, most high-powered computer, so don't let sales hype sway you into buying more computer than you need.

Activities that require higher levels of speed and performance include:

- Working with sophisticated graphics when performing tasks such as photo manipulation or web design.
- Working with audio and video.
- Uploading and downloading larger files over a network using a more recent Wi-Fi specification.
- Rich multimedia experiences such as gaming.

Consider how much time you will spend on the computer. If you work at home and will use the computer eight hours a day, five days a week, you need a better quality (which often means more expensive) model. But if you only log on to read email a few days a week and manage your checkbook with a financial program once a month, a lower-end model will probably do. Efficient power consumption is also a consideration. Many systems comply with Energy Star requirements, but double-check the labeling and specifications to make sure.

Also consider how your choices create a larger impact on the world around you. Think about how the rapid change and constant upgrading of electronic devices has an impact on the environment. Technology doesn't seem green when you think about all the computer monitors, cases, keyboards, hard drives, and printers that get disposed of each day. The green computing movement focuses on several important goals:

- Reduce hazardous materials.
- Make computing devices energy efficient.
- Recycle outdated computing devices.
- Limit factory waste.

Technology companies have been developing more energy-efficient models with longer lasting batteries and promoting recycling efforts to reduce the impact of discarded equipment on the environment. Solid state drives that contain fewer moving parts and consume less power are also a growing trend.

What Processor Speed Do You Need?

As you learned earlier in this chapter, computers contain a processor that is located on a computer chip. Your computer **processor speed** influences how fast your computer runs programs and completes various tasks. **Clock speed** relates to the speed at which a processor can execute computer instructions.

Both clock speed and processor speed are measured in **gigahertz (GHz)**. The more gigahertz, the faster the speed. Processor speed gets faster all the time. As of the writing of this textbook, 3 to 3.4 GHz was the higher end for the average computer, though the 3.6 GHz quad-core processor reaches the equivalent of 14.4 GHz with a single-core processor.

Those who want to play games and work with graphics require a more powerful computer system.

Gordon Moore, one of the founders of Intel, is the creator of **Moore's Law**, which states that over time the number of transistors that can be placed on a chip will increase exponentially, with a corresponding increase in processing speed and memory capacity. Since proposing this idea in 1970, Moore has largely been proved correct.

How Much Memory and Storage Is Enough?

Your computer has a certain amount of **memory capacity** that it uses to run programs and store data. When you buy a computer, you'll notice specifications for the amount of RAM and hard drive storage each model offers.

- As you read previously, RAM is the memory your computer uses to access and run programs. RAM chips come in different types, including DRAM, SRAM, and SDRAM. Most modern computer memory is some variation of SDRAM, including DDR-SDRAM, DDR2-SDRAM, DDR3-SDRAM, and DDR4-SDRAM. Performance has steadily improved with each successive generation. The more sophisticated programs you run and the more you want to run several programs at one time, the higher RAM you should look for.
- RAM chips are rated by **access speed**. This rating measures how quickly a request for data from your system is completed. Your computer will also use some RAM to run the operating system and application programs. RAM access speed is measured in **megahertz (MHz)**. For example, 800 MHz is a typical access speed that would be sufficient to run most computers. Chip manufacturers continue to work to develop new, faster processors. Adding RAM to your computer is simple to do and one of the least expensive ways to improve the speed of your computer.
- The average computer hard drive's capacity for data storage is measured in **gigabytes (GB)**. A typical consumer PC might have anywhere from 500 GB to 750 GB of storage capacity, though terabyte systems measured in thou-

sands of gigabytes are becoming common. Large systems used by business and research institutes may have storage measured in petabytes (PB), or one quadrillion bytes. Likewise, data storage size is an issue for smartphones and tablets. Many smartphones and tablets feature up to 128 GB of storage.

Which Operating System Is Right for You?

Chapter 4 of this book goes into operating systems in detail, but when you go out shopping for a computer, there are a few reasons you should consider which operating system to use.

All new computers come with an operating system installed, so the cost of the operating system (OS) isn't usually a factor. However, if you are buying OS software on its own (for example, to upgrade to a newer version), there are variations in cost to consider.

Windows is a popular operating system, but if you are buying the software (as opposed to buying a computer with it already installed), it is costly. Also, because of its popularity, Windows computers are more often the target of viruses (though Macs are gaining in both popularity and virus problems).

Linux is a Windows-like operating system that comes in different "flavors" such as Mint and Ubuntu. You can use the free, open source version of Linux or you can purchase a packaged edition. For a packaged edition, the company may charge a fee and add something extra, such as support and documentation. The Linux community offers lots of applications and add-ins to choose from.

Mac computers are manufactured by Apple and use the macOS operating system. While Apple offers its own software written by Apple or third-party Apple developers, many software applications originally written for Windows are also available in Mac versions, such as Microsoft Office. You can also set up your Mac to run the Windows operating system alongside macOS, so you can take advantage of a wide variety of software.

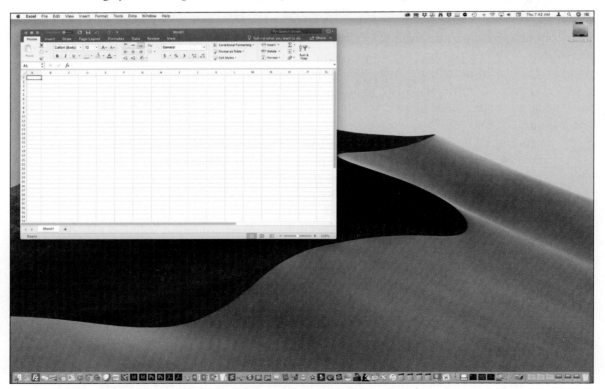

Operating systems are getting better at allowing users to interact with files created on other systems.

What Are You Willing to Spend?

Computers range in price from a few hundred dollars to several thousand. Often, buying a base model and customizing it with additional memory and an upgraded monitor will get you the system you need. Laptops are still slightly more expensive than desktop models, but lightweight laptops such as ultrabooks have leveled the playing field. It's a good idea to use online sites to check out the best prices and read consumer reviews before buying.

Where you shop can have an impact on price. You can shop for a computer in an online store, in a traditional retail store, or you can use online auctions or classifieds to find deals on new or refurbished models. Retail stores provide the ability to try before you buy. You can spend time with the computer, getting the feel of the keyboard and viewing the display to see if it meets your needs. You can also direct any questions you have about the computer to a real, live salesperson. Manufacturer sites allow you to customize computers to your requirements. To help you answer your specific questions, many online sites provide shopping support using real-time chat, which can make buying online easy.

Check to see whether memberships in organizations such as your university or member-discount retail stores such as Costco give you access to good deals. Also, think about shipping costs if you buy online, and note that retail stores often charge a restocking fee if you return a computer.

Should You Opt for Higher-End Graphics and Sound?

Those who work with a lot of visual elements (for example, photographers, gamers, or movie buffs) should opt for a better quality graphics card when purchasing a computer. Movies and games also use sound, so a high-end sound card is a plus too.

Computers that have higher-end sound and image capabilities are referred to as *gaming* or *multimedia models*. In addition to more sophisticated sound and video cards, they usually have higher memory specifications.

Companies such as Dell and Lenovo offer buyers the opportunity to build a custom computer online.

Course Content
Take the Next Step Activity

Review and Assessment

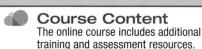
Summing Up

3.1 A World of Digital Devices

Some digital devices have multiple features built into one unit, such as a laptop computer or cell phone. Others use other **peripheral devices** that physically or wirelessly connect with them.

3.2 The Parts That Make Up Your Computer

The **motherboard** is the primary circuit board on your computer and holds the central processing unit (CPU), BIOS, memory, and so on. The motherboard is a container where the various working pieces of your computer slot into one compact package. Circuits on the motherboard connect the various components contained on it.

The central processing unit, which is a microprocessor, sits on the motherboard. The CPU processes a user's requests. A **multicore processor** contains more than one CPU, which in this situation is referred to as a *core*. Things that affect the processing speed of your computer include **clock speed**, cache, and the number of cores.

The motherboard also holds different types of memory. **Read-only memory (ROM)** is the permanent, or nonvolatile, memory of a computer, hardwired into a chip. In a **PC**, the **BIOS** is stored in ROM. BIOS stands for basic input/output system, and this code is embedded in a memory chip also residing on the motherboard. During the boot-up sequence, the BIOS checks devices to ensure they are working properly and then starts them up. It also directs the hard drive to boot up and load the operating system (OS) to memory. Random access memory (RAM) chips store data and programs and allow access in any order, which speeds up processing.

While some computers have built-in sound or graphics cards, others include them as **expansion cards**.

A **power supply** in a desktop or laptop computer is located where the power cord is inserted at the back of the system unit. The power supply switches alternating current (AC) provided from your wall outlet to lower voltages in the form of direct current (DC). **Wireless charging** uses an electromagnetic field to charge a device.

A computer uses a **port** to connect with a peripheral device such as a monitor or printer or to connect to a network. A **serial port** is used to connect a peripheral device to the serial bus via a cable with a plug that typically contains nine pins. A **universal serial bus (USB) port** is a small rectangular slot that provides a popular way to attach peripherals and storage devices. You can add a hub to expand the number of available USB ports. A **FireWire port** is based on the same serial bus architecture as a USB port. FireWire provides a high-speed serial interface for peripheral devices that require high performance, such as digital cameras, camcorders, or external hard disk drives. A **Thunderbolt port** is a peripheral connection standard that supports high-resolution displays and other devices that require a lot of bandwidth, such as high-end video cameras or data storage devices. An **Infrared Data Association (IrDA) port** allows you to transfer data from one device to another using infrared light waves.

Some technologies, such as Bluetooth, allow for wireless communications between your computer and other devices.

All storage media have methods for reading the data from the media (input) and for writing the data to the media (output). The various storage media used on a computer are accessed using **drives**.

Storage media include the **hard disk** built into your computer; an **optical drive** to read a **CD**, **DVD**, or **Blu-ray disc**; an **external hard drive** or a **network attached storage (NAS)** shared external hard disk; a **flash drive**; and a **solid-state drive (SSD)**. Third-party cloud storage systems store and manage data on a web-connected server.

3.3 Input and Output Devices

You use **input devices** to get data into your computer and **output devices** to get information out.

In addition to a **keyboard** and a **mouse**, computer input devices include **touchscreens**, **touchpads**, **scanners**, digital cameras and **webcams**, **gaming devices**, and **microphones**. Specialty input devices include foldable keyboards for **mobile internet devices (MIDs)**, **assistive technology** devices, and devices often used in retail or manufacturing settings, such as **bar code readers** and **RFID readers**.

Output devices include any device that displays, prints, or plays content stored in your computer, including your **monitor**, **speakers**, **Bluetooth headsets**, **printer**, or an **LCD projector**. **4K** and **HDR** provide the latest resolution quality for monitors.

Some devices provide both input and output functionality, such as a **fax machine**.

3-D printing is a method of using printers to create 3-D objects from digital models. A 3-D printer prints layers on top of one another to build a physical object, and 3-D printing is increasing in popularity as the technology develops.

3.4 Purchasing a Computer

When buying a computer, you should consider the following questions:
- What are your computing needs?
- What **processor speed** do you need?
- How much memory and storage is enough?
- Which operating system is right for you?
- What are you willing to spend?
- Should you opt for higher-end graphics and sound?

Your computer has a certain amount of **memory capacity** that it uses to run programs and store data. When you buy a computer you'll notice specifications for the amount of RAM and hard drive storage each model offers. RAM **access speed** is measured in **megahertz (MHz)**. The average computer's hard drive capacity for data storage and the storage capacity for smartphones and tablets is measured in **gigabytes (GB)**.

All new computers come with an operating system installed, so consider whether you want Windows, Linux, or a macOS system.

Computers range in price from a few hundred dollars to several thousand. Often, buying a base model and customizing it with additional memory and an upgraded monitor will get you the system you need. Laptops are still slightly more expensive than desktop models. Where you shop can have an impact on price.

Those who work with a lot of visual elements (for example, photographers, gamers, or movie buffs) should opt for a better quality graphics card. Movies and games also use sound, so a high-end sound card is a plus.

Terms to Know

3.1 A World of Digital Devices
peripheral device, 60

3.2 The Parts That Make Up Your Computer
motherboard, 61

chip, 61

multicore processor, 61

parallelized, 62

read-only memory (ROM), 62

nonvolatile memory, 62

BIOS, 62

expansion card, 62

PC card, 62

power supply, 62

charging mat, 63

wireless charging, 63

port, 63

physical port, 63

serial port, 63

universal serial bus (USB) port, 63

FireWire port, 64

Thunderbolt port, 64

Infrared Data Association (IrDA) port, 64

MIDI, 64

wireless adapter, 65

drive, 65

hard disk, 65

CD, 65

DVD, 65

optical drive, 65

Blu-ray disc, 65

optoelectronic sensor, 65

external hard drive, 65

network attached storage (NAS), 65

flash drive, 66

flash memory, 66

solid-state drive (SSD), 66

3.3 Input and Output Devices
input device, 67

output device, 67

keyboard, 67

mouse, 68

infrared (IR) technology, 68

touchscreen, 68

stylus, 68

digital pen, 68

touchpad, 68

scanner, 68

webcam, 68

gaming device, 69

wired data glove, 69

microphone, 69

mobile internet device (MID), 69

assistive technology, 69

bar code reader, 70

RFID reader, 70

keystroke logging software, 70

monitor, 70

speaker, 70

Bluetooth headset, 70

4K, 70

high dynamic range (HDR), 70

thin film transistor active matrix liquid crystal display (TFT active matrix LCD), 71

light-emitting diode (LED) display, 71

surface-conduction electron-emitter display (SED), 71

organic light-emitting diode (OLED), 71

printer, 72

photo printer, 72

thermal printer, 72

plotter, 72

fax machine, 72

3-D printing, 72

LCD projector, 72

document camera, 73

interactive whiteboard (IWB), 73

virtual reality system, 73

augmented reality, 73

3.4 Purchasing a Computer

Projects

Check with your instructor for the preferred method to submit completed work.

Understanding Your Computer System and Performance

Project 3A

Knowing the parts of your computer system will be helpful when you need to upgrade your system or buy a new computer. Using the following list of computer parts as a guide, create a table and include descriptions of the features of your current computer. If you don't own a computer, use a public computer or a friend's computer to perform this task. Note any hardware features that you want to upgrade for increased performance. (You can find information about the computer system in the Control Panel of Windows under the category of System and Maintenance; under System Preferences, General on a Mac; or by looking at your user manual or visiting your manufacturer's website and searching for your computer model.) Be prepared to discuss your findings in class.

CPU processor	Monitor type	Ports (types and numbers of each)
CPU speed	Monitor size	
RAM	Sound card	Size
Hard drive capacity	Video card memory	Weight
DVD drive	Speakers	Style (desktop/laptop/ other)

Project 3B   TEAM

Even though technology has improved and computers are more reliable than they were several years ago, they can still have problems. In assigned teams, search for two computer diagnostic tools that can help to resolve a potential computer issue. Compare and contrast the tools and summarize your findings. Include in your summary information about what the software will provide for solving issues related to hardware devices, operating systems and installed software, hardware, speed, performance benchmarks, and system stability. Write a memo to your instructor discussing the advantages of using a diagnostic software utility to identify problems and enhance the performance of your computer.

Purchasing a Computer

Project 3C

You have decided to purchase a new computer to launch a home business. Because your current computer works fine, you would like to donate it to a charitable organization. Before making the donation, you know that your computer must be wiped clean of data. You realize that simply deleting files or reformatting your hard drive doesn't remove all the data, so you decide to research the proper way to permanently erase computer data. Based on your research, prepare a list of instructions for wiping your computer clean of data.

Project 3D TEAM

Your Introduction to Computers class has volunteered to help people in the community who need to buy new computers. Your instructor will divide the class into three groups:

- Group A is purchasing a computer for a senior citizen for personal use.
- Group B is purchasing a computer for a teenaged gamer.
- Group C is purchasing a computer for a business person who travels regularly and needs internet access and basic business functionality.

Your group's task is to meet the needs of your assigned community member by researching a variety of computer options. To begin the process, consider the user's likely computer needs and cost limitations. Then investigate different computer systems that satisfy those requirements. Use the table format created in Project 3A to guide you in your research, and keep a list of your reference sources. Prepare a group presentation that shares your research findings and computer recommendation.

Exploring Data Storage

Project 3E

With an increased need for computer users to access information from any location, online data storage is becoming a popular service. Research several cloud storage companies, and examine the pros and cons of the service, the costs, and any security issues. Write a report about online data storage and include a recommendation for an online storage company. Be sure to provide your instructor with a list of references for your information.

Project 3F TEAM

As a help desk member of an information technology department, you observe some risky data handling practices. In particular, you have noticed that many employees have discs, flash drives, and external hard drives lying around on their desks. You have also observed some employees placing discs and flash drives in their pockets to carry them home, and posting files in the cloud to access while on the road. After you speak with your supervisor about the potential for damaged or lost data, your supervisor asks you to work with other members of the help desk staff to prepare a slide presentation about the proper use of data storage devices. As a team, select three storage devices and provide guidelines that address how to keep the devices and the data stored on them safe. Be prepared to present these guidelines to your class, and to submit a list of references to your instructor.

How Much RAM Is Enough?

Project 3G

Your classmate is having a hard time understanding RAM. To help your classmate, you have been asked by your instructor to research RAM and to create an outline covering its basic concepts. Your outline should include the definition of RAM, the types of RAM, the location of RAM, and the amount of RAM a well-equipped computer used by an average college student should contain. Keep a list of references that you used in your investigation. After you have completed the outline, work with a partner to review each other's outlines and to offer feedback on their usefulness. Record any comments on each other's outlines, and submit your outline and list of references to your instructor.

Project 3H TEAM

A local school has received a donation of computers that are in excellent shape, but they are operating slowly and need a new software suite installed. A school employee, Tony, has volunteered to install the software. Tony realizes that the computers need more RAM to run the new software and address the slow operating speed, but he has never installed RAM. To help Tony with this task, prepare a tutorial presentation (supported with text and screen captures) on the installation of RAM. Be sure to include a list of references for your information.

Wiki Project—Exploring New Computer Technologies

Project 3I

Research the internet to find an article about a new computer device released in the current calendar year. If possible, focus your search on devices that use a new technology that wasn't available a year ago. Read the article and write a summary, including a description of the device, its purpose and functions, the total cost of the device (including all components), and where you can buy it. Attach a list of references that you used to write your summary, and post the summary on the course wiki site.

Project 3J TEAM

You or your team will be assigned to edit and verify the content posted on the wiki site from Project 3I. If you add or edit any content, make sure you include a notation within the page that includes your name or team members' names and the date you edited the content—for example, "Edited by [student name or team members' names] on [date]." Keep a list of references that you used to verify your content changes. When you are finished with your verification, include a notation at the end of the entry—for example, "Verified by [student name or team members' names] on [date]."

Green Computing

Project 3K

The increasing use of computer technology has had a significant impact on the environment. As a result, many companies have implemented policies that support green computing. Research the term "green computing" and the major initiatives behind this environmental movement. Write a report that discusses the definition and main concepts of green computing. Address the measures being taken by companies to support this movement and explain their positive impact on the environment. Be sure to offer a recommendation in your report as to how computer users can support the green computing movement. Submit your report and a list of your references to your instructor.

Project 3L **TEAM**

Your information technology team has been assigned to prepare a slide presentation on green computing for the board of directors of your school. This presentation should provide the broad definition and purpose of green computing as well as specifically define and address the major concepts of this environmental initiative: green use, green design, green disposal, and green manufacturing. Be sure to include in your presentation a recommended green computing plan for your school to implement.

Class Conversations

Topic 3A What happens to old computing devices?

The personal computer began to become a household fixture in the early 1990s. In the last 25 years, computer manufacturers have increased productivity and built computers that are faster, smaller, and better year after year. What happens to all the old computers, tablets, and smartphones? What impact will these old devices have on our environment in the next 10 years?

Topic 3B What happens when drones deliver?

In the UK, Amazon used drones to deliver an Amazon Fire TV and a bag of popcorn. From the time the order was placed, delivery with the drone took 13 minutes. Other companies are exploring the use of drones to make delivery fast and efficient. For example, Google delivered Chipotle at Virginia Tech and Domino's delivered pizza in New Zealand, both via drone. What impact will this type of technology have on our economy? Will flying drones affect our quality of life? What are some other possible uses of drones?

Topic 3C How can you shop smart for a computer?

Computers are sold in a variety of settings, such as specialty computer stores, large discount retailers, and online retail stores and auction sites. What are the benefits of purchasing a computer at a specialty computer store? How can you tell if a particular online store is secure when you are making a purchase from it? Is the support you can get for your computer better if you buy it at a brick-and-mortar store? What support options do you think exist for a computer purchased online?

System Software
The Control Center of Your Computer

What You'll Accomplish

When you finish this chapter, you'll be able to:

4.1 List tasks performed by system software, describe the steps performed by the operating system in starting your computer, and outline the history of operating system development.

4.2 Explain platform dependency and differentiate the popular operating system packages in use today.

4.3 Describe tasks the operating system performs, list system maintenance utilities that are included in an operating system package, and explain how to send the computer to sleep or shut it off.

4.4 Explain the differences between an operating system for a PC and an operating system for a tablet or smartphone and list the common mobile operating systems in use today.

Why Does It Matter ⑦

Hardware without system software is like a powerful new car without a driver. The computer may have the latest gadgetry under the hood, but without system software, the gadgetry can't do anything. Every computer, from a small mobile device to a large mainframe computer, needs system software, which includes the operating system and various utility programs, to work. Understanding how an operating system package such as Windows manages the various devices, programs, and files on your computer will help you make choices when purchasing a computer, troubleshoot your system when things go wrong, keep your data secure, and perform regular maintenance to keep your computer running well.

 The online course includes additional training and assessment resources.

System software performs basic computing functions, including starting your computer and loading the operating system, which provides the interface between you and the machine. It also provides tools for configuring and maintaining your computer system and managing programs and files.

Course Content

Take a Survey

Technology in Your Future Video

4.1 What Controls Your Computer?

The first instructions your computer uses when you turn on the power are stored in a ROM BIOS chip located on the motherboard (the circuit board that holds the various elements of your computer). These instructions are called *firmware*. **Firmware** in a computer system contains code that is used to start the computer and load system software.

System Software

System software includes your operating system and several types of utility software. The **operating system (OS)** provides you with an interface to work with your computer hardware and applications, while the **utility software**, or utilities, optimizes and maintains your computer. The OS and utilities are typically combined in an operating system package such as Windows, Linux, UNIX, macOS, Android, or Chrome OS. Portable devices such as tablets and smartphones typically use a mobile operating system that provides similar functionality to the OS on your desktop or laptop computer.

The OS part of the system software allows you to organize and control your computer hardware and software. It's in charge of loading files, deciding which applications get to do what and when as you work, and shutting down your computer. The OS is essential for you to interact with your computer because software and hardware simply can't run without an OS in place. Your OS translates your commands and performs appropriate actions.

The utilities included in an operating system package such as Windows aren't *essential* to the functioning of the OS but are *useful* to the OS and to the user. **Disk Cleanup** in Windows, for example, helps the OS function by maintaining your computer and getting rid of unused or unusable files.

Figure 4.1 shows some of the devices and components that are managed by the OS.

Starting Your Computer

The process of starting your computer, called **booting**, is handled by instructions that reside on the BIOS (basic input/output system) chip slotted onto the motherboard. These instructions load the OS, which then loads the remaining system and software files into RAM (memory). If you start a computer when the power is turned off, you're performing a **cold boot**; if you restart the computer (shut it down and then turn it on again without turning the power off), you're performing a **warm boot**. A new specification for booting your computer, called **UEFI (Unified Extensible Firmware Interface)** is meant to eventually replace the aging BIOS firmware and

FIGURE 4.1 Examples of Devices and Components Run by the OS

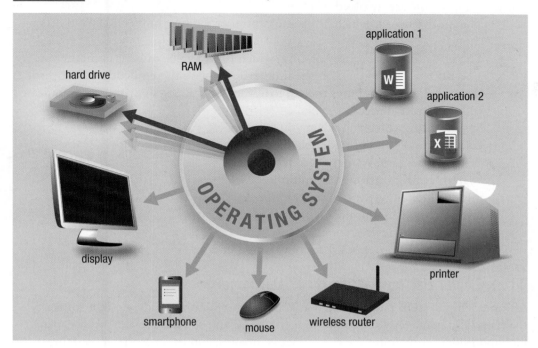

Though outsold by Windows, macOS has a loyal and growing following. This is the macOS desktop.

makes booting computers a much speedier process. UEFI provides a layer of security called *Secure Boot* that prevents unauthorized code from being installed on your computer. UEFI can stop malware from loading before your operating system and its security features are running.

Figure 4.2 shows the steps in booting a computer. During this procedure, the following components are involved:

- The motherboard, which holds the central processing unit (CPU) and other chips.
- The CPU, or microprocessor chip, which is the brains of your computer.
- The BIOS or UEFI chip on the motherboard with the embedded code (firmware) that your computer uses to load the operating system and communicate with hardware devices.
- **System files**, which run when you start up your computer and provide the instructions that the operating system needs to run.
- **System configuration**, a definition or means of defining your entire computing system, including the identity of your computer, the devices connected to it, and some essential processes that your computer runs.

> " Rebooting is a wonder drug—it fixes almost everything. "
> —Garrett Hazel, "Help Desk Blues"

The Operating System Package

An **operating system package** such as Windows or Linux includes system software that runs and manages your computer's hardware and software resources; organizes files and folders containing your documents, images, and other kinds of content; and helps you perform maintenance and repairs when your computer has problems. Operating system packages also offer security features such as password protection to keep others from using your computer and a firewall to prevent someone from remotely accessing your computer.

Operating system packages also include basic applications you can use to get your work done or be entertained, such as simple word processing programs like WordPad;

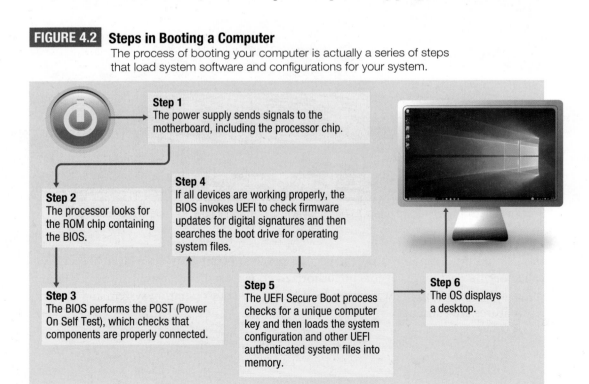

FIGURE 4.2 **Steps in Booting a Computer**
The process of booting your computer is actually a series of steps that load system software and configurations for your system.

Step 1
The power supply sends signals to the motherboard, including the processor chip.

Step 2
The processor looks for the ROM chip containing the BIOS.

Step 3
The BIOS performs the POST (Power On Self Test), which checks that components are properly connected.

Step 4
If all devices are working properly, the BIOS invokes UEFI to check firmware updates for digital signatures and then searches the boot drive for operating system files.

Step 5
The UEFI Secure Boot process checks for a unique computer key and then loads the system configuration and other UEFI authenticated system files into memory.

Step 6
The OS displays a desktop.

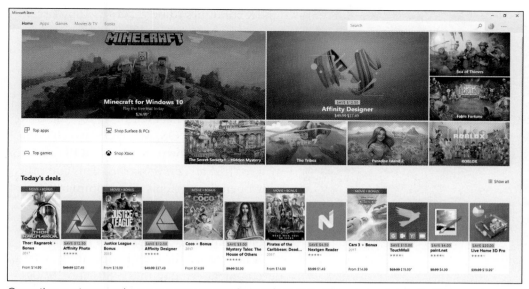

Operating systems make your computer work, but they also offer access to applications and games to entertain you. For example, the Windows Store app is where you can download apps and games. App stores offer new content often.

games like Spider Solitaire; media players such as Windows Media Player to play music or videos; and tools such as a calculator, a calendar or address book, and an internet browser.

A (Brief!) History of Operating Systems

There was a time when computers had no operating system. In this pre-OS time, every program had to have all the required **drivers** (software that allows an operating system to interface with hardware) and specifications needed to connect to hardware such as printers. In the early days of computing, functions in software were simple enough that this system worked. As programs became more sophisticated and hardware grew more complex, something was needed to orchestrate the interaction between software and hardware.

Mainframe computers—the pre-consumer computers that were often the size of a large desk or van—typically used an operating system created for them by their manufacturer. UNIVAC I, the first commercial computer produced in the US, is an example. In the pre-personal computer period, names of the operating systems were not significant because the hardware drove the purchase decision.

UNIX, an OS written with the C programming language, was developed by AT&T Bell Laboratories in the 1970s and became popular with corporations for running their workstations. It was usable across a variety of hardware and could be customized by the companies who licensed it.

In 1951, the Remington Rand Corporation introduced the first commercial computer produced in the US, the UNIVAC I.

DOS sported white, amber, or green letters on a black background and required that the user memorize commands to use it.

UNIX continues to be widely used by larger organizations today, and there are several versions of this OS.

With the development of microprocessors, small personal computers began to catch on, and by the 1980s it was clear a more standardized OS was needed. In 1980, Microsoft produced its first OS, MS-DOS, which eventually became known simply as **DOS** (an acronym for *disk operating system*). This and other early operating systems were **command-line interfaces**, meaning that you typed commands as text, but they were not in plain English—and they weren't very intuitive.

A huge breakthrough in operating system development came in the 1980s and 1990s when the **graphical user interface (GUI)** was introduced to the public first by Apple and subsequently used by Microsoft. A graphical user interface was a much more user-friendly way to work with a computer, because you could click icons and choose options from menus and dialog boxes, rather than typing commands.

All subsequent versions of Windows incorporated a graphical user interface. With Windows 8 and Windows 8.1 Microsoft expanded and improved touchscreen functionality in its operating system, representing a significant shift in how users provide input to their computing devices. With the release of Windows 8 in 2012, the design of Windows was modeled after mobile device operating systems. Windows 10 improved and expanded the design to provide a consistent interface across all types of Windows 10 devices.

> **Course Content**
> Take the Next Step Activity
> Ethics and Technology Blog

4.2 Perusing the Popular Operating System Packages

The major operating systems used by individual consumers in the world today are Windows; macOS for Apple computers; and Linux, a free OS used on PCs by people who prefer a free, customizable OS alternative to Windows. Smartphone and tablet devices use mobile operating systems such as the Apple iOS and Linux-based Android. All operating systems can take advantage of computing "in the cloud."

Understanding Platforms and Platform Dependency

The term **platform** is used to describe the combination of hardware architecture and software used to run applications. The platform's OS may be dependent on the hardware. For example, a Mac operating system isn't designed to run on a Windows computer. This is called **platform dependency**. Table 4.1 lists five common operating system platforms.

Our Digital World

TABLE 4.1 Operating System Platforms

Platform	Operating Systems
PC/Windows	Windows 10, Windows 7
Mac/Mac OS and macOS	OS X in various versions: El Capitan, Yosemite, Mavericks, Lion, Snow Leopard. The first version of macOS was High Sierra; Mojave was released in 2018.
Tablets	Windows 10, Android, Chrome OS, iOS
PC to Mainframe/UNIX	New versions of UNIX, including SchilIX, Belenix, and AIX, are now released as open source software through the Open Solaris project.
PC to Mainframe/Linux	Linux, being freely distributable, comes in many different distribution "flavors" such as Red Hat, SUSE, and Ubuntu, as well as a wealth of versions for specific languages such as Chinese (Sunwah Linux) and Norwegian (Skolelinux).

Multitasking demands that the OS handle several functions at once.

However, platform dependency is becoming less of an issue as consumers demand cross-functionality when they buy computers and software. Today, software such as Boot Camp, which is included with macOS, allows you to run Windows on a Mac. Though there are still different versions of productivity software, such as Word for PCs, Word for Mac, and Apache OpenOffice Writer, you can typically save files in each version that are compatible with (able to be used by) any other platform.

Still, when you purchase application software such as a computer game or word processor, you should make sure that you buy programs that will work with the platform on which your computer is based. If you are using software in the cloud, the cloud service provides the latest version compatible with your computer.

> "It does not matter which browser, operating system, program or service you use—it falls down to whether the job can be done or not. If it can, then continue. If not, by all means move on."
>
> —Zack Whittaker, ZDNet

Today's popular graphical operating systems: Windows 10, macOS Mojave, and Linux Ubuntu.

Our Digital World

Windows **Microsoft Windows** was initially a graphical interface layered on top of the DOS operating system. Eventually, it became a true GUI OS with a robust feature set, and today, Windows is the most widely used operating system in the world. It can be used on computers from a variety of manufacturers, and there are thousands of software products available to run on Windows. Regular updates keep the Windows OS current as issues become known or new security fixes are required. Windows provides several ways to customize the interface, including changing the background and colors, adding tiles, and customizing the Start menu.

The latest version of Windows OS is **Windows 10**, which provides an intelligent assistant and search tool named **Cortana**. Users can interact with their desktop, laptop, or tablet computer using a touchscreen; and on the **Start menu**, live tiles that represent apps such as Weather, People, and Finance update on the fly.

Several features in Windows 10 help you find and do your work more easily, such as Timeline, which provides a list of recent activities. The ability to sync devices to pick up activities where you left off on any device allows you to keep track of your activities across devices. In addition, Near Share facilitates sophisticated file sharing; Focus Assist mutes alerts when you need to focus on your work; and Cortana's Organizer makes it easier to view reminders and lists.

An update in the fall of 2018 included a new Your Phone app designed to make an Android device work better with Windows. For example, you can use the app to copy a photo stored on your phone and it paste into a document. You can also use the app on your PC to send a text message that is routed through your phone. File Explorer had a new dark mode theme applied in the same update.

Mac Those who swear by Macs and the Mac operating system, currently **macOS**, have several reasons to do so. Macs have sophisticated graphics-handling capabilities and clear, crisp screens. Their computer designs are unique and include models such as the new iMac Pro, boasting a 27-inch LED-backlit 5K display with integrated CPU, and the sleek MacBook, weighing in at just over 2 pounds.

In their early days, Mac computers were heavily marketed to the educational world, though PCs have since made significant inroads in that environment. Creative industries such as graphic design and photo imaging have also been historically strong markets for Macs because of the computers' ability to handle advanced graphics-related tasks. Today, the popularity of iPhones, iPads, and the iTunes online store have convinced many more people to adopt Apple's hardware, software, and services. At the time of writing, the macOS version of Mojave added features to help you manage your files and to use your computer in tandem with your iPhone. The Finder feature was updated to allow you to easily find, tag, and even change files without launching an app. Apple also made some of their most popular iOS apps available for Macs.

> " This planet came with a set of instructions, but we seem to have misplaced them. Civilization needs a new operating system. "
>
> —Paul Hawken, Environmentalist

With macOS Mojave, Apple introduced their new look called *Dark Mode*. Dark Mode is designed to showcase the colors and details in your content with toolbars and menus in muted colors. The new Stacks feature help you declutter your desktop by grouping related files.

Mac's increased popularity comes with a downside, however. Once thought to be a less virus-prone system, the macOS has been receiving more attention from programmers of malicious software. For example, in January 2018, Apple confirmed that Macs, iPhones, and iPads were affected by the Meltdown and Spectre malware that allowed hackers to steal data from unsuspecting users, and in October 2018, Apple confirmed that malware was discovered on one of their servers. Although Mac devices are still less susceptible to certain types of malware, Mac users must practice malware prevention just like their Windows counterparts. Luckily, antivirus programs can combat malicious software on both Mac and Windows systems.

UNIX **UNIX** is primarily a server operating system. Created by a handful of AT&T Bell Labs employees, UNIX was designed to run servers that support many users. UNIX uses a command-line interface and, because it is written in the C programming language, it is more portable across platforms—meaning that it can run on all types of computers, including PCs and Macs. UNIX is a popular choice for web servers that support thousands or millions of users. Several other server operating systems exist, such as Open Enterprise Server, as do many other specialized server operating systems, such as those used to run web applications and email.

Linux **Linux**, first developed by Linus Torvalds in 1991, is, to a great extent, based on UNIX. In fact, some people believe it may eventually take over UNIX's market. Linux can be used as either a network operating system or a personal computing OS. Linux is an **open source** operating system, meaning that its source code is freely available for anyone to use or modify. Modifications contributed over the years have resulted in various distributions or versions of Linux, such as Red Hat, Ubuntu, Mandriva, and SUSE.

Linux can run on multiple platforms, and users report that the OS is highly stable and flexible. When it first appeared, Linux was seen as the rebel's OS and was supported by people who were against Microsoft and the general dependency on their products. Though Linux hasn't toppled Windows, it has definitely found its place in the OS world.

With the macOS desktop (shown here in Dark Mode), apps and settings can be launched using the menu system, icons on the dock along the bottom of the screen, or in a window.

Linux is not just for personal computers and servers. Many people know **Android** (a modified version of Linux) as the operating system on their smartphone or tablet. Various flavors of Android are also used as operating systems on PCs. In 2016, Remix OS, an Android operating system for PC, was released. Sporting a desktop similar in appearance to that of Windows 10, including a taskbar, Remix OS is now discontinued. Other editions of Android for PC, including Phoenix OS and Android x86, are available. Proponents of Android on a PC OS favor using apps and games on a larger screen with mouse and keyboard support, as well as taking advantage of multitasking ability.

The Linux logo is a penguin, which has been known to appear in many costumes and versions.

Other computing environments, such as mainframe computers and supercomputers used by larger organizations and governments, also require an operating system. These operating systems are often based on Linux or UNIX and are highly customized to suit the hardware architecture and the special purposes that the computers are designed to serve. For example, IBM uses z/OS on many of its established mainframe systems.

Network Operating Systems

Computer networks have unique requirements to share resources, such as printers, among users. They must also continually respond to numerous requests from each connected device. For example, computer networks have to support both network servers and users' computers (called *clients*)—processing several hundred or thousand requests at a time. They also have to make sure that each request is routed to the correct location and is answered promptly and accurately.

Network operating systems typically have to handle several tasks at the same time, referred to as **multitasking**. Because people don't like to be kept waiting, a network OS needs to be able to process and prioritize many user requests at once.

Network operating systems also have to interface with all computers, regardless of the client's computing platform. By supporting Windows, Macs, and Linux-based computers, for example, a network OS enables them all to share data and resources.

In addition to handling all those users, a network operating system also has to keep data secure while it's being transmitted or stored. It does that using a system of **least possible privileges** to manage user access to resources and data, which dictates that each user gets only those privileges he needs to get his work done.

Finally, a network OS must interface with other networks that may be using a different network operating system and applications specific to that system. For example, a request might come in from a Windows-based network to a Linux network. The Linux network may retrieve the requested data and send it back to the Windows network, where it is then loaded into a Unix system for use in a web application that shares data over the internet.

Phoenix OS, a version of Android for PC, has similarities to Windows 10 but closer inspection reveals a different menu structure and the use of icons rather than tiles.

The Future of the OS

It is possible that operating system packages could become less important, or at least less visible, to the average user in the not-too-distant future.

Google's **Chrome OS** is a Linux-based operating system meant to appeal to people who work primarily in web-based applications. In June 2011, Chromebooks—PCs running the Chrome operating system—manufactured by Samsung and Acer were released, and in 2013, Toshiba, HP, and Google entered the market, expanding the Chromebook choices. In 2017 Google announced that their Chrome OS would allow Android apps to run natively, laying the groundwork for the first Chrome OS tablet to be released in 2018. In late 2017, Google also entered the Chromebook market with the release of the Pixelbook: a device with 2,400 x 1,600 pixel display and 7th-generation Intel Core 'i' processor at an affordable price.

Chromebooks boast an eight second startup time with all settings, applications, and documents stored in the cloud. Using the web to deliver the majority of the user experience makes Google Chrome a cross-platform OS, because it works on any computer that supports browsers (which is essentially all computing devices in use today, including smartphones).

Google's Chrome OS provides a web-based experience.

While the Chrome OS is currently only available on the Chromebook platform, other new consumer-oriented operating systems offer cross-platform functionality. While originally an operating system for mobile phones, Google's Android is also offered as an operating system on tablets and some computers. Similarly, the **iOS** from Apple powers some iPod models, iPhones, and iPads.

In a sense, Chrome OS, Android, and iOS all function as web operating systems, providing access to services online. Android and iOS also allow users to customize their experience by downloading and installing apps on a device, while Chrome OS offers web apps. These modern operating systems allow consumers to take advantage of a variety of mobile computing platforms and customize them to work wherever and whenever they are needed. With its Windows 10 operating system, Microsoft is betting that a touchscreen approach to input and a continuously updated OS will be the wave of the future.

> **Course Content**
> Take the Next Step Activity
> Ethics and Technology Blog

4.3 The Tasks of the Operating System Package

Although there are different operating system packages out there, they have similar basic functions in common. The following sections provide a rundown of those functions and what they help you do with your computer.

Providing a User Interface

A **user interface** is what you see when you look at your computer screen. Most operating systems provide a graphical user interface, which means that icons and pictures are used to represent the text-based commands used in operating systems of the past.

Your computer's **desktop** is like home base for your computer, from which you use a text or graphical interface or menus to run applications and work with files. As illustrated in Figure 4.3, GUIs use graphical tiles, buttons, and panes or windows that display operating system settings or open files. You can open more than one window at a time; for example, one window may show your web browser at the same time that another window displays a photo editing program. You can also customize your Start screen and/or desktop to use a different background image or color, or change the color scheme for all of your screen elements, including window borders and title bars.

Configuring Hardware

As previously defined, a driver is a small software program that provides the instructions the OS needs to communicate with a piece of hardware, such as a printer or keyboard. You may notice when you first plug in a new device, such as a wireless mouse receiver into a USB drive, that the OS has to find the driver before you can use the mouse. If your system has the wrong driver for a device, you may have problems; for example, a printer may print garbage text if you're not using the correct driver.

FIGURE 4.3 **Some Customizable Settings for Windows PCs**
Operating systems allow you to customize their environment with different backgrounds, resolution settings, and tools.

Tap or click a category to display that category's customizable options in the right pane.

Change the lock screen image by tapping or clicking the tile for the desired image or the Browse button to select a picture stored on your device.

Click or drag the slider to the *On* position to play a slide show when the screen is locked.

Most operating systems today come packaged with many common device drivers or offer a way for you to download a driver from the web. Most devices today work with **Plug and Play**, a feature that recognizes devices you plug into your computer—for example, into a USB port. Once the Plug and Play feature identifies the hardware, the OS can install the necessary driver, if the driver is available.

Most operating systems don't require a CD to install drivers for peripheral hardware, Rather, you can download drivers from the web.

You may have to update drivers now and then—especially when you upgrade your OS. Your OS may be set up to perform regular system updates that download and install newer drivers for hardware automatically, and the hardware manufacturer may alert you to updated drivers if you have registered your computer.

Controlling Input and Output Devices

Your operating system also controls input and output devices. Commonly used input devices include your keyboard, mouse, stylus on a touchscreen, joystick (used for gaming), and microphone. These devices provide a way to interact with your computer, telling it what to do through typing, clicking, speaking, or selecting options in menus and dialog boxes (Figure 4.4). In addition, input devices (typically your keyboard) allow you to enter text into documents you create using programs such as word processors or spreadsheet software. Output devices include your monitor, printer, and speakers. These generate visual, printed, and audio information from your computer.

Input and output devices may be wired, which means you connect to them by plugging them into your computer, typically through a USB port; or they may be connected using various wireless options, leaving your input device free from restrictive wires. Your operating system uses device drivers to set up and control input devices. Each input and output device has a unique driver.

Some computing devices, such as tablets and smartphones, allow you to use a touchscreen to communicate with the OS. Your finger or a stylus becomes the input device, rather than a mouse or keyboard.

Managing System Performance and Memory

The speed with which your computer functions, called its **performance**, is largely determined by the computer processor, available cache, bus, and the amount of memory installed in the computer. All of these are orchestrated by system software.

Utilities in your operating system package monitor system performance and provide information about system resources. This allows you to troubleshoot performance, adjust settings, change which programs the OS loads when you start up, and so on.

The following is a brief rundown of the four most significant factors controlling system performance.

FIGURE 4.4 **Microsoft Word Input Options**
The most common input options for interacting with your operating system and applications are a touchscreen, the keyboard, and the mouse.

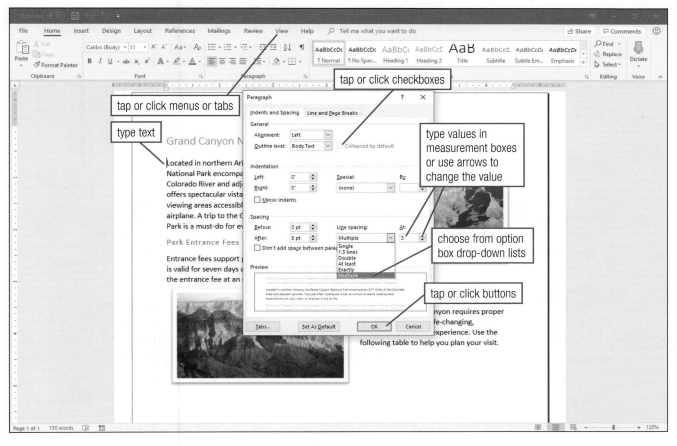

♡ Device performance & health

Check that your Windows is up-to-date and if there are any issues impacting your device health. The Health report shows the status of the most recent scan.

🗒 Health report
Last scan: 3/22/2018

⊘ Windows Update
No issues

⊘ Storage capacity
No issues

⊘ Device driver
No issues

⊘ Battery life
No issues

⊡ Fresh start

Start fresh with a clean and up-to-date installation of Windows. This will keep your personal files and some Windows settings, and remove some of your apps.

In some cases, this may improve your device's startup and shutdown experience, memory usage, Store apps performance, browsing experience, and battery life.

Additional info

System performance monitoring tools on Windows 10 (top) and Mac (bottom) computers.

Activity Monitor (My Processes)

CPU | Memory | Energy | Disk | Network | Q Search

Process Name	% CPU	CPU Time	Threads	Idle Wake Ups	PID	User
lsd	7.1	1:16.98	4	1	425	nancymuir
screencapture	5.2	0.22	3	0	33061	nancymuir
Activity Monitor	2.5	1.06	5	2	33057	nancymuir
General	0.6	0.80	5	0	33010	nancymuir
UserEventAgent	0.2	3:23.80	27	0	418	nancymuir
https://support.apple.com	0.2	2.89	6	8	33037	nancymuir
gamed	0.1	3:18.26	4	1	668	nancymuir
com.apple.preference.notific...	0.1	0.65	5	0	33008	nancymuir
sharingd	0.1	5:24.00	6	0	476	nancymuir
Spotlight	0.0	1:51.21	7	1	897	nancymuir
identityservicesd	0.0	1:42.29	7	0	437	nancymuir
Safari	0.0	4:54.74	7	1	14185	nancymuir
CommCenter	0.0	19:52.23	8	1	423	nancymuir
cfprefsd	0.0	1:19.64	6	0	16260	nancymuir
System Preferences	0.0	10.26	4	0	33000	nancymuir
quicklookd	0.0	0.15	5	0	32987	nancymuir
Safari Networking	0.0	18:00.58	8	1	14187	nancymuir
Finder	0.0	3:37.30	8	0	870	nancymuir
distnoted	0.0	3:02.92	3	2	420	nancymuir
com.apple.CommerceKit.Tran...	0.0	0.17	4	1	33058	nancymuir
Dock	0.0	6:00.83	3	0	868	nancymuir
SafariCloudHistoryPushAgent	0.0	5:43.25	3	0	95350	nancymuir
SymUIAgent.app	0.0	26:15.90	19	0	925	nancymuir

System:	4.19%	CPU LOAD	Threads	1908
User:	6.87%		Processes:	320
Idle:	88.94%			

Memory Memory refers to the capacity for storage in a computer. One kind of memory is permanent—for example, read-only memory (ROM), which holds information such as the BIOS and start-up instructions for the operating system. Another kind of memory is temporary—for example, random access memory (RAM), which stores data while your computer is operating but loses that data when you shut down your computer. **Virtual memory** is the part of your operating system that handles data that cannot fit into RAM when you are running several programs at once. Figure 4.5 shows how, when RAM is used up, data is stored or "swapped" into virtual memory (the file moved into virtual memory is called a **swap file**). As each program is loaded and data fills RAM's "bucket," overflow data spills over into virtual memory.

Cache Cache memory is a dedicated holding area in which the data and instructions most recently called from RAM by the processor are temporarily stored. The process of caching allows your computer to hold a small amount of recently used data in memory so that, when you ask for something again, the OS doesn't have to run around in RAM looking for it. This is similar to the function of RAM, but because the cache sits on the microprocessor, it can be accessed more quickly than RAM, which is stored on another chip. The larger the cache size, the more data can be held and quickly retrieved, though a cache always has a maximum size to keep the process of searching through it efficient.

Processor The CPU, sometimes referred to as the computer's *processor*, is an electronic circuit that runs your computer's hardware and software. The OS coordinates the use of a processor's resources by scheduling and prioritizing tasks. In systems with more than one processor core (dual- or quad-core systems), the OS manages tasks between or among processors. A dual-core system is shown in Figure 4.6.

Parallel processing describes a computer operation in which several calculations or program instructions are carried out simultaneously. On a PC with two or more cores on a single CPU chip, or two or more CPUs, the OS allocates the work to be done amongst the cores or processors, an action referred to as **multiprocessing**.

Bus A **bus** is a subsystem that moves instructions and data among the components in your computer (for example, between RAM and the processor and between the processor and the cache). The faster the bus, the faster the computer.

FIGURE 4.5 **Using Up RAM**
RAM is used up quickly, and virtual memory supplements RAM.

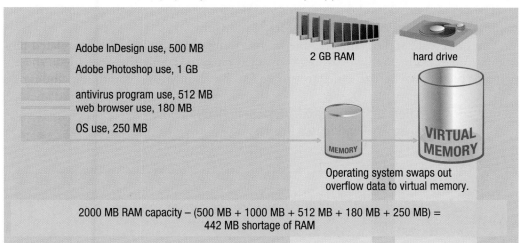

Adobe InDesign use, 500 MB

Adobe Photoshop use, 1 GB

antivirus program use, 512 MB
web browser use, 180 MB

OS use, 250 MB

2 GB RAM

hard drive

MEMORY

VIRTUAL MEMORY

Operating system swaps out overflow data to virtual memory.

2000 MB RAM capacity – (500 MB + 1000 MB + 512 MB + 180 MB + 250 MB) = 442 MB shortage of RAM

FIGURE 4.6 **Sending a Task to the Processor**
In a dual processor configuration, the OS determines which processor completes a task.

SAVE

PROCESSOR 1

PROCESSOR 2

One measurement of bus speed is how much data can move at one time. Another is the speed at which the data can travel. A bus with a larger capacity to move data at a faster speed helps your computer complete tasks more efficiently.

Routing Data between Applications and Devices

It's midnight, and you've just finished a lengthy homework assignment and want to print a copy to submit to your instructor the next morning. You click the Print button in your word processor program and grab the pages as they come out of the printer. Ever wonder what your operating system is doing in the background?

Figure 4.7 shows how this routing of data between an application and a printer works and what role the OS plays. Your OS may also route data to and from your network or other devices such as a scanner.

Managing File Systems

The operating system provides the features used for storing and retrieving files. The OS has to keep track of the physical location where a document is saved on your hard disk; to do this, it maintains the **file allocation table (FAT)**. Because the bits that make up a file may be stored all around your hard disk and not in a contiguous group, this table provides an index of data locations the OS can reference, helping to speed up the time it takes to open a file.

The OS also provides the commands you use for naming, organizing, and maintaining files, such as Rename, Delete, Move, and Copy.

You can create and organize your own hierarchy of file folders to store related files in one spot. This hierarchy provides a so-called **path** to the file that starts with the name of the drive where the files are stored. All drives in your computer, including

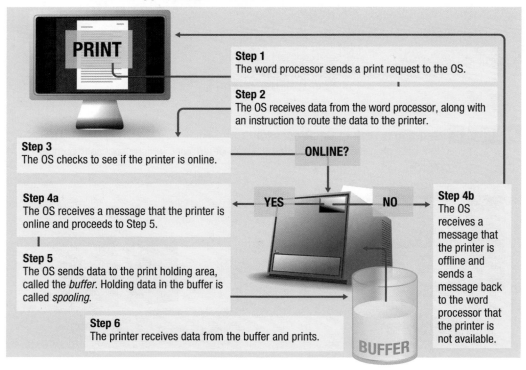

FIGURE 4.7 **The Role of the Operating System in Routing Data from an Application to a Device**

PRINT

Step 1
The word processor sends a print request to the OS.

Step 2
The OS receives data from the word processor, along with an instruction to route the data to the printer.

Step 3
The OS checks to see if the printer is online.

ONLINE?

Step 4a
The OS receives a message that the printer is online and proceeds to Step 5.

YES

NO

Step 4b
The OS receives a message that the printer is offline and sends a message back to the word processor that the printer is not available.

Step 5
The OS sends data to the print holding area, called the *buffer*. Holding data in the buffer is called *spooling*.

Step 6
The printer receives data from the buffer and prints.

BUFFER

Our Digital World

your hard drive, USB drives, CD/DVD drive, and network drives, are designated with drive letters on Windows-based computers. The macOS Finder feature uses a similar structure.

For example, a word processing file named *Acton Engineering Invoice.docx* might be filed in your Documents folder (Figure 4.8). The file location might have a path of This PC → Documents → Invoices → Acton Engineering Invoice, where

- *This PC* is the entire contents of your computer and any external storage devices that are attached.
- *Documents* is the documents folder created for your user account name.
- *Invoices* is the subfolder created by a user to help organize the Documents folder.
- *Acton Engineering Invoice* is the word processing file.

Providing Search and Help Capabilities

All operating systems provide a method of searching your computer for the files you need and a help system you can use to look for information about how to use your computer.

Search features allow you to enter information about a file, such as the date it was last saved, a word contained in the file name, or the actual contents of the file, and the operating system helps you locate the file. Features such as Finder on a Mac and File Explorer on a Windows computer also help you search through hierarchies of folders and files to find what you need.

FIGURE 4.8 **Windows File Management**
The path name, shown in File Explorer, identifies the location where you stored a document.

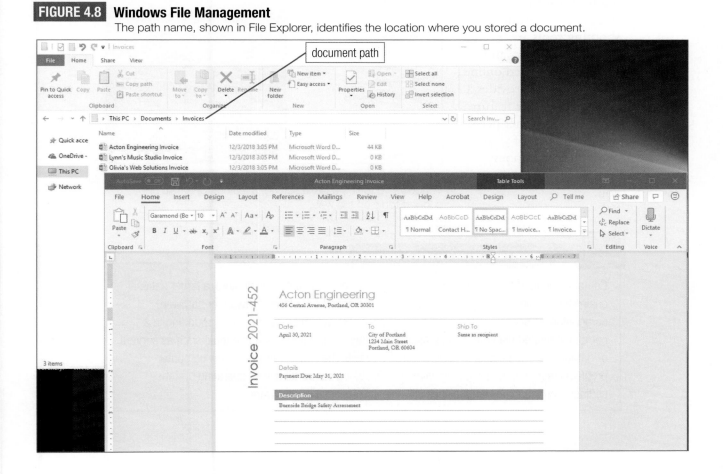

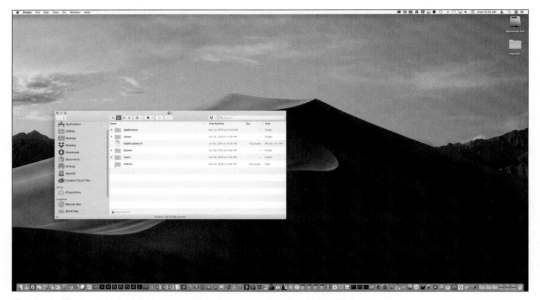

The Mac Finder window helps you locate files in a hierarchy of folders.

Help features provide searchable support information and troubleshooting tools for the operating system. Some troubleshooting tools are located on your computer and others are accessed online. Windows even provides a remote assistance feature that allows another person to take control of your computer to pinpoint your problem and fix it for you.

You can use the Cortana search box on the taskbar in Windows 10 to find a file, a setting, a feature, or a help and support topic on the web.

Computers in Your Career

Computer support specialists may work in a small, local computer repair store, a larger national chain store, or an expansive technical support call center for a large software or hardware corporation. To become a computer support specialist, you must have a solid grounding in a computer-related field (at least an associate's degree or certification in a topic such as servers or databases) and be able to communicate well with the public. Because many computer problems require troubleshooting within the OS, being comfortable with operating system settings, configurations, and utilities will help you succeed in your work.

Maintaining the Computer with Operating System Utilities

System maintenance is an important task for system software. This process is similar to taking your car into the shop on a regular basis for tune-ups to keep it running efficiently. There are several utility programs included in your operating system package that you can use to perform system maintenance tasks, scan drives and files for problems, and otherwise help protect your system and troubleshoot problems. Typical utility program functions include those listed in Table 4.2.

TABLE 4.2 Utility Program Features

Function	Windows Feature Name	Mac Feature Name
Backing up or restoring damaged files	File History (Windows 10); Backup and Restore (Windows 7)	Time Machine
Performing automatic updates to get updated drivers or definitions	Windows Update	Software Update
Cleaning up your hard drive to get rid of temporary files or files you haven't used for a long time	Disk Cleanup	Repair Disk
Defragmenting your hard drive so that scattered bits of files are reorganized for efficiency in retrieving data	Optimize Drives (Windows 10); Disk Defragmenter (Windows 7)	
Restoring your system to an earlier date to remove new settings that may have caused problems	System Restore	Time Machine
Scheduling regular maintenance tasks such as backing up and downloading updates	Task Scheduler	Time Machine
Uninstalling unwanted programs	Uninstall or Change a Program	Click app and press Delete
Providing security features such as a firewall or spyware protection	Windows Firewall and Windows Defender	Systems Preferences, Security & Privacy

Sending the Computer to Sleep or Shutting It Off

The operating system is in charge of shutting down your computer. The way you shut down your computer is important, because following the proper procedure prevents the loss of unsaved data and properly saves system settings. In addition to shutting down your computer, you can use features with names such as Sleep or Hibernate to save computer power but still be able to return to your desktop and any running programs quickly, without having to reboot.

Optimizing the power options for your computer or mobile device is an important strategy for green computing. Laptops and mobile devices are preset to enter a power saving state after a period of inactivity. You can shorten or lengthen the time before a power saving mode kicks in. If you're not going to use the computer for a while, the Shut Down option ensures power consumption will be zero. Conserving power on your computing devices will not only save you money on your electricity bill but will also contribute to a greener environment.

Course Content
Take the Next Step Activity

4.4 Going on the Road with Mobile Operating Systems

A **mobile operating system** (also known as a mobile OS or mobile platform) is the brains behind mobile **smartphones** such as the Samsung Galaxy, Google Pixel 2, Sony Xperia, and iPhone. These devices contain a rich feature set that essentially makes them into small computers.

What Makes a Mobile OS Different?

A mobile OS, which is stored in ROM, is not nearly as robust as a personal computer operating system, because a phone simply doesn't have a great deal of memory, and most users don't need as much functionality from their phones as they do from their personal computers.

A mobile OS also uses and accesses smaller resources so it can run quickly and efficiently, with an emphasis on conserving battery life. The mobile OS also manages touchscreen functionality and wireless broadband connections.

A Wealth of Mobile OS Choices

Stroll through an electronics store and you'll notice dozens of mobile phone models, including smartphones that include an operating system. Different phone manufacturers adopt different operating systems for their various smartphones. Each mobile OS also has an online store where users can download applications (called *apps*) for their mobile phones.

Smartphones come with several popular apps already installed. However, people love downloading new apps as they discover various ways to use their smartphones. Applications that users download to their phones typically only work with that phone's OS. If you buy a new phone with a different operating system, you'll need to buy new apps.

According to StatCounter Global Stats, the top two mobile operating systems worldwide as of September 2018 are Android (77%) and Apple's iOS (21%). The remaining 2% of the market is split among smaller, niche mobile operating systems. The market for mobile devices varies by geographic region and can be changed rapidly by innovation. What is popular today may be outdated a year from now. Table 4.3 lists today's most common mobile operating systems.

The family of mobile telecommunications standards has gone through a progression, reaching its fourth generation, called **4G**. The first generation used analog signals to transmit sound. Generation 2 made the leap from analog to digital, and **3G** turned mobile phones into mini-computers that could send and receive information

TABLE 4.3	Common Mobile Operating Systems
OS	**Description**
Android	An open source, cross-platform OS supported by Google used by many manufacturers
iOS	Apple-developed iPhone and iPad OS based on Mac OS X
Windows 10 Mobile	Microsoft's mobile OS based on Windows 10
Ubuntu Touch	An open source OS based on Linux

faster than ever. 4G does what 3G did but offers higher data rates, greater reliability, and improved security, which allows service providers to support several improvements in telecommunications.

In 2018, telecommunications providers rolled out their first 5G networks. Verizon offered 5G service in four markets in 2018 with plans for more deployments in 2019. **5G** involves a new industry standard for radio signals that will see wireless connectivity speed 30 to 50 times faster than 4G connectivity, with improved reliability. 5G smartphones are expected to be widely available by the end of the decade, but smartphones aren't the only devices 5G is designed to improve. 5G networks are also being built to accommodate the future of IoT, in which devices will communicate with one another constantly.

With so many different mobile operating systems, the variety of options and incompatibilities between them can prove confusing and inconvenient for consumers to manage and difficult for wireless companies to support.

The widespread use of mobile devices such as tablets has resulted in employees bringing their own devices into the workplace and connecting the devices to company networks. This is referred to as **BYOD (bring your own device)** and is also used to refer to an employee connecting his or her personally owned smartphone or laptop to the company network.

> " [BYOD is] one of the single most important steps in motivating business productivity. "
>
> —Adriana Karaboutis, CIO, Dell

A wealth of mobile operating systems run on phones from different manufacturers.

A wearable device such as a smartwatch, a fitness tracker, or smart glasses needs an operating system too. These devices come in a range of sizes and capabilities. The OS used in these devices operates and controls many of the same resources that a smartphone does, but within a smaller environment and without sacrificing performance and battery life. This is an emerging field that is changing rapidly. Table 4.4 lists three common operating systems used in wearable devices sold today.

TABLE 4.4 **Common Wearable Device Operating Systems**

OS	Description
Wear OS	The Google-supported open source operating system used in some smartwatches and other fitness trackers
Tizen Wear	An open source OS residing with the Linux Foundation for use in the popular Samsung Gear and Neo smartwatches
watchOS	Apple's mobile OS (based on iOS) specifically designed for use in its smart watch

Computers in Your Career

Web designers that specialize in responsive web design are in demand. When people started browsing the web from their tablets and smartphones, web designers' initial strategy was to create a mobile version of each website with content adjusted to minimize the need to expand the display or pan from side to side. Eventually, web designers realized a responsive design, which adjusts the website layout based on the device's screen size, was a better solution. In this approach, a web developer creates code that allows a website to query each device's screen size and then adjust the web content accordingly. For example, on a restaurant site, menus are just as readable on a 4-inch smartphone screen as they are on a 27-inch desktop monitor. Successful responsive web designers have knowledge and experience in the latest versions of JavaScript, HTML, and CSS, as well as media queries.

Computers in Your Career

When new versions of an operating system or application appear and companies upgrade, there are a few key jobs to be handled. If you provide network support in a corporation, you will help to *roll out* new versions of software to all necessary employees by installing and customizing the software. If you are a trainer, you may be called upon to provide instruction in the new software features. Other IT (information technology) workers may run a help desk, answering technical questions and troubleshooting problems that come up in the first few weeks or months of use.

Course Content
Take the Next Step Activity

Summing Up

4.1 What Controls Your Computer?

System software contains an **operating system (OS)**, which you use to interact with hardware and software, and **utility software** to maintain your computer and its performance. An operating system package—for example, Windows 10—contains the operating system plus certain utilities.

Starting your computer, called **booting**, is carried out by instructions located on a chip on your computer's motherboard. If you start by turning the power on, you're performing a **cold boot**. If you restart the computer (shut it down and then turn it on again without turning the power off), you're performing a **warm boot**.

Early computer manufacturers used their own operating systems. Microsoft's first OS, MS-DOS, which eventually became known simply as **DOS**, was a **command-line interface**, meaning that you typed in text commands.

A huge breakthrough in operating systems happened in the 1980s and 1990s when the **graphical user interface (GUI)** was introduced by Apple and then by Microsoft, creating a much more user-friendly way to work with a computer.

4.2 Perusing the Popular Operating System Packages

The term **platform** identifies the architecture of a computer and the OS intended to run on it. **Platform dependency** is becoming less and less important as consumers demand cross-functionality when they buy computers and software.

The major operating systems used by individual consumers today are **Microsoft Windows** for PCs; **macOS** for Apple computers; and various versions of **Linux**, an **open source** operating system. **UNIX** is an OS used by larger organizations to run their networks. **Android** (a modified version of Linux) is also used as an operating system on some PCs.

Network operating systems require the ability to control access among user computers (called *clients*) by assigning the **least possible privileges**. They must also work with multiple other operating systems while keeping users secure.

Other operating systems such as Google's **Chrome OS**, Android, and **iOS** allow for computing on smaller devices such as tablets and smartphones and enable the user to buy apps, either online through a browser or by downloading to the device.

4.3 The Tasks of the Operating System Package

Operating system packages have certain basic functions in common:

- Providing a **user interface**, including a **desktop**.
- Configuring hardware using device drivers.
- Controlling input and output devices.
- Determining the speed at which the computer functions (its **performance**), which involves the processor, available cache, bus, and memory.

- Processing data. **Parallel processing** describes a type of computer operation where several calculations or program instructions are carried out simultaneously. On a PC with two or more cores on a single CPU chip, the OS allocates the work to be done amongst the cores or processors—an action referred to as **multiprocessing**.

- Providing information about system resources to enable the user to troubleshoot performance, adjust settings, change which programs the OS loads upon startup, and so on.

- Routing data between applications and devices—for example, sending a document to a printer or receiving an image from a scanner.

- Managing systems and folders to store and retrieve information.

- Providing search capabilities for locating files and a help system for information about how to use your computer.

- Performing system maintenance tasks through utility software included in your operating system package.

- Shutting down the system or putting the system into a power-saving state.

4.4 Going on the Road with Mobile Operating Systems

Mobile operating systems are used to power **smartphones** (phones with a rich feature set that emulate some of the power of computers) and tablets. A mobile OS is stored in ROM and uses smaller resources to make sure that it runs quickly and efficiently.

Terms to Know

4.1 What Controls Your Computer?

firmware, 86
system software, 86
operating system (OS), 86
utility software, 86
Disk Cleanup, 86
booting, 86
cold boot, 86
warm boot, 86
UEFI (Unified Extensible Firmware Interface), 86

system files, 88
system configuration, 88
operating system package, 88
driver, 89
DOS, 90
command-line interface, 90
graphical user interface (GUI), 90

4.2 Perusing the Popular Operating System Packages

platform, 90
platform dependency, 90
Microsoft Windows, 93
Windows 10, 93
Cortana, 93
Start menu, 93
macOS, 93
UNIX, 94

Linux, 94
open source, 94
Android, 95
multitasking, 95
least possible privileges, 95
Chrome OS, 96
iOS, 97

4.3 The Tasks of the Operating System Package

4.4 Going on the Road with Mobile Operating Systems

Projects

Check with your instructor for the preferred method to submit completed work.

Customizing the Desktop

Project 4A

You help out at your aunt's internet café. She has asked you to come up with a customized desktop for two PCs she provides for patrons who don't have their own laptop or tablet. She wants her café logo to display on the desktop background instead of the standard Windows desktop image. She also wants you to create a guest account on the PC for customers who want to browse the internet. Before you tackle this job, you need to learn how to customize a Windows PC by changing the background image and color scheme. You also need to investigate how to create a user account that doesn't require a Microsoft account to sign in.

- Research how to modify the background of your desktop, change the color scheme, and create a user account called a *local account*.

- Change the background and color scheme of your desktop. *Note: Some computer labs in schools are configured to prevent access to the personalization features of the desktop. If necessary, do this activity on your personal computer.*

- Open a window to show your new color scheme. Resize the window to take up approximately one-third of the screen.

- Move the window to the bottom right edge of the screen so that as much of the new background as possible is visible.

- Capture an image of your revised desktop using the Print Screen key. Paste the image into a blank Microsoft Word document. Below the image of the screen, type the steps that you completed to change the background and color scheme.

- Add another section to the document that lists the steps you used to create a local account.

Project 4B `TEAM`

Many companies install a desktop PC at each employee's desk and employ technology tools that prevent the employees from customizing their PCs. These tools generally prevent employees from changing some settings on their PCs, including installing new programs and deleting existing ones. Consider reasons why a company would use these technology tools to ensure that each PC is standardized. What types of workplaces would encourage customization and allow employees to put their personal stamp on the desktop? Prepare a slide presentation with the reasons for standardization and the types of workplaces your team suggests might allow customization.

Mobile Mania

Project 4C

Your school's alumni committee has assigned you the task of purchasing a smartphone complete with a one-year service contract that includes unlimited talk, text, and data to be given as a door prize at the alumni dinner. You have narrowed down the choices to the latest versions of the Samsung Galaxy and the Apple iPhone. Research these smartphones on the internet, and prepare a table listing the pros and cons of each phone. Aside from the benefits and drawbacks, consider the costs associated with purchasing the phone along with the one-year contract. After a careful review of all factors, write a summary of your research including your recommendation for the door prize.

Project 4D TEAM

The manager of the school's alumni office has asked your team to prepare a list of items that will be given to a mobile app programming team to create a new app for the school's alumni office. The free app will be available from the Google Play Store and the Apple App Store. To help you get started, the manager has already said she wants alumni to be able to access news and upcoming alumni event information. Research your school's alumni association to see what activities and/or benefits the association offers alumni. If possible, interview a few alumni to find out what they would want in a mobile app. Brainstorm with your team the types of activities that a user would want the new app to provide. Prepare a brief presentation for the alumni office manager with your team's recommendation for the mobile app programming team.

Housekeeping 101

Project 4E

Using utility programs to perform system maintenance tasks is important to the efficient operation of a computer, yet many people neglect these tasks. For example, we have all lost a file or know of someone who has lost a file on his computer. If all of us maintained daily backups, losing data from a computer malfunction or user error would not be a problem because we could simply restore the data from our backup copy.

Consider five maintenance tasks that every computer user should perform on a regular basis to avoid data loss and keep his or her computer running efficiently. Create a survey checklist with your five maintenance tasks. Next, choose a representative sampling of 10 friends and/or relatives who use a computer for a variety of functions. Provide them with a copy of your checklist and ask each person to check off the tasks he or she performs routinely. Prepare a spreadsheet that displays the survey data results, identifying the tasks that are and are not routinely performed. Write a summary with anecdotal information about data losses from your survey participants as they filled out your checklist. In your summary, write your conclusion as to the importance of computer maintenance among your survey participants. Provide an explanation as to whether the survey is or is not reasonably representative of the general population of computer users.

Project 4F TEAM

Read the article at https://ODW5.ParadigmEducation.com/ComputerChecklist for a list of 29 tasks a technical support technician performs to keep a computer running smoothly. (As you are reading, keep in mind that this list is geared toward computer maintenance in a corporate setting by a dedicated IT staff.) Then, within your team, discuss the tasks on this list that would be useful for home computer users, and select the 10 most important tasks. As a group, design a Top 10 Maintenance Task Checklist that could be printed and distributed to home computer users. For each task on the checklist, note the frequency with which the task should be performed—for example, weekly, monthly, or annually. Also, be sure to include a tip for each task that provides users with the location of the system tools needed within the operating system software.

Accessibility Options

Project 4G

You work as a volunteer with senior citizens at a local community center. These individuals have asked for your assistance with customizing their settings to make their computers more user-friendly. Specifically, the seniors have mentioned difficulties with reading small type, seeing the mouse pointer, and hearing system sounds during certain activities. Research various Windows Ease of Access options. Experiment with various settings until you find a group of Ease of Access options that you think will address the seniors' needs. Prepare a handout with instructions on how to set up the Ease of Access options you have decided to recommend. Include screenshots where appropriate.

Project 4H TEAM

Many senior citizens at the community center have arthritis, making it difficult for them to type on a computer. To help them with this task, you would like to set up speech recognition on their computers. Research the speech recognition technology that is included with the operating system. Determine the key features of the program and whether any special hardware is needed for the program to work. Also, find out if the speech recognition feature has support for Spanish, the first language of several of the senior citizens. Prepare a presentation that summarizes the results of your research and evaluates the adequacy of the program for the needs of these individuals.

Wiki Project—File Management Tips and Tricks

Project 4I TEAM

You or your team will be assigned to address one or more of the following file management questions:

- What is the distinction between a file and a folder? How is each one visually represented?
- What are the default folders provided with the operating system?
- What is a file extension? Should file extensions be displayed?

- How do I browse the folders on my computer or other storage device?

- How can I search for a file?

- How do I perform common file management tasks?

- What can I do with deleted files that have been sent to the Recycle Bin (Windows) or Trash (Mac)?

- What are the pros and cons of using cloud storage (such as OneDrive) for all my data?

Post to the course wiki site your response for the topic assigned to you. Write the post as if it were going to be used as a help document for people unfamiliar with file management.

Project 4J TEAM

You or your team will be assigned to edit and verify the content posted on the wiki site from Project 4I. If you add or edit any content, make sure you include a notation within the page that includes your name or team members' names and the date you edited the content—for example, "Edited by [student name or team members' names] on [date]." Keep a list of references that you used to verify your content changes. When you are finished with your verification, include a notation at the end of the entry—for example, "Verified by [student name or team members' names] on [date]."

Wiki Project—Troubleshooting Tips

Project 4K TEAM

You or your team will be assigned to one or more of the following topics related to troubleshooting computer problems:

- Help! My printer is not working.

- Help! My internet is not working.

- Help! My file will not open.

- Help! My computer keeps freezing up.

- Help! I need to create a PDF of my document and I don't know how.

- Help! No sound is coming out of the speakers.

- Help! I don't know how to attach a file to an email message.

- Help! I don't know how to share pictures between my smartphone and my computer.

Post to the course wiki site your response for the topic assigned to you. The goal is to provide the average computer user with simple, precise instructions to follow to resolve the problem or answer the question without calling in an expert. When writing your instructions, use clear, concise language and avoid technical jargon that might confuse your audience.

Project 4L TEAM

You or your team will be assigned to edit and verify the content posted on the wiki site from Project 4K. If you add or edit any content, make sure you include a notation within the page that includes your name or team members' names and the date you edited the content—for example, "Edited by [student name or team members' names] on [date]." Keep a list of references that you used to verify your content changes. When you are finished with your verification, include a notation at the end of the entry—for example, "Verified by [student name or team members' names] on [date]."

Class Conversations

Topic 4A   What happens if the mouse and keyboard go away?

With the popularity of digital assistants such as Alexa and Google Home and the increased use of voice recognition on our smartphones and other smart appliances, can you envision a time when all users will interact with their computer through touch-screen and voice without using a mouse or keyboard? Consider how the workplace might need to be redesigned to accommodate everyone talking to their computers. Also think about your daily activities and how talking to a device might have an impact on you or your family members when you're out in public. How will you handle privacy concerns when you are in a public place or a place where others can hear you?

Topic 4B   What more should a smartphone do?

With 4G smartphones and tablets becoming more commonplace, many consumers are used to pulling out their smartphones to watch a movie or television show, video message a friend, or video call a friend or relative across the world using Skype or FaceTime. Thousands of apps are available for smartphones for just about any purpose. What features should the next generation (5G) of smartphones provide that offer functionality you lack with your current smartphone?

Topic 4C   How is a file naming standard developed?

A large amount of the time that a user spends interacting with the OS involves file management tasks. Browsing the content of a USB drive, a hard drive, or your cloud storage account, and looking for that elusive file that you created last week, the name of which you can no longer remember are common tasks. A file name that seemed clever and memorable last week will no longer be memorable three months down the road. The searchability of an OS has improved with each new release, but one can still spend an inordinate amount of time looking for a file. Assume you have been hired by a company to devise a standard file naming system for all employees to use when naming new files. Consider how you would go about developing such a standard. For example, what questions would you need to ask end users before you attempted to develop a set of "rules" to use when assigning new file names? Consider elements that should be included in file names (such as names or revision dates) for workplaces where employees save different versions of files to shared network folders. How would the file folder hierarchy be developed so that each department has similar folders and subfolders? What challenges do you foresee in developing a file naming standard?

Chapter 5

Application Software
The Key to Digital Productivity

What You'll Accomplish

When you finish this chapter, you'll be able to:

5.1 State the role of application software.

5.2 Discuss the role of major categories of application software with examples of products in each category.

5.3 Describe how software is created, obtained, and priced.

5.4 Explain how software products can use content created in other software products.

Why Does It Matter ?

Your computer exists largely to run application software—the software you use to get things done. In your personal life and on the job you will use software that helps you accomplish tasks, and you will learn how to use new programs to help you complete your work more efficiently. Understanding the different types of software that exist and how they are used will help you take advantage of a wide variety of tools to increase your productivity.

The online course includes additional training and assessment resources.

We use application software to get our work done, from writing and calculating to analyzing and presenting information. But we also use application software for fun, to play games, listen to music, view photos, and read books. We can use software on a computer, tablet, or even a mobile phone.

Course Content
Take a Survey

Technology in Your Future Video

5.1 Software's Role in the World of Computing

System software, specifically the operating systems described in Chapter 4, enables your computer to function and run **application software**. Application software helps you do many things such as get your work done, learn something new, communicate with others, create art, or play games. Most software is available for desktop and mobile devices including laptops, tablets, and smartphones.

Application software started out providing basic functions. Products such as VisiCalc, the first spreadsheet application for personal computers, were basically glorified calculators. Early word processors, such as WordStar, offered little more than the ability to enter and edit text to create simple documents. When using these products, you had to key command codes to enter or delete text and even to scroll (move up and down the page) through a document.

In the next section, you will discover what major categories of software exist today and how each category is used.

With primitive interfaces and functionality, early software such as VisiCalc (right) and WordStar (below) provided the foundation for today's software applications.

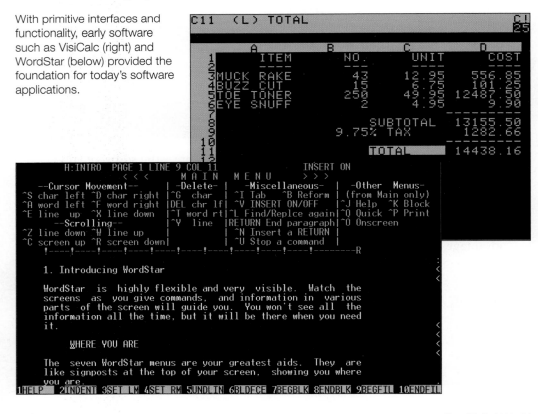

5.2 The Many Types of Application Software

From the early days of consumer software, which consisted mainly of basic business tools and simple games, software application development has exploded for computers and all varieties of mobile devices. A huge number of categories of software applications now exist, each with sophisticated feature sets. Using application software, you might:

- Produce business documents such as reports, memos, budgets, charts, presentations, product catalogs, and customer lists.
- Manage massive amounts of data for student, patient, or employee records.
- Complete a learning exercise, a course, or even an entire degree without ever entering a classroom.
- Create works of art, including photos, drawings, animations, videos, and music.
- Organize content such as photos, appointments, contacts, home inventories, and music playlists.
- Manage your personal finances, deposit and withdraw money from a bank, handle your tax reporting, or generate legal documents.
- Play games or pursue hobbies such as sports or interior design.
- Stream music or video for entertainment.
- Communicate with others individually or in large online conferences.
- Perform maintenance or security tasks that help keep your computer functioning and your data secure.

In fact, software is so much a part of our lives that it's difficult to categorize modern application software products by work or personal use. For instance, word processors aren't just used for business, and design software isn't just used by artists. Financial software is used for creating department budgets and invoices as well as to track personal finances.

In this section we have categorized software applications by what they help you do. As you read through the following sections, consider how many categories of software you use.

Productivity Software

Productivity software includes software that people typically use to get work done, such as a word processor (working with words), spreadsheet (working with data, numbers, calculations, and charting), database (organizing and retrieving data records), or presentation program (creating slideshows with text and graphics). Software suites such as Microsoft Office, Apache OpenOffice, and Apple iWork combine productivity applications into one product, because many people use two or more of these programs to get their work done. Software suites often include a word processor, a spreadsheet application, presentation software, and database management software. Suites also allow users to easily integrate content from one program into another, such as including a spreadsheet in a report created with a word processor. Productivity software and suites are also available for many smartphones and other mobile devices so you can be productive from anywhere. Productivity suites available in the cloud—notably Office 365 and Google Docs—and the ability to store data files in the cloud and sync between devices enable you to work wherever you have internet connectivity.

Word Processor Software **Word processor software** certainly does "process" words, but today it does a great deal more. With a word processor such as Microsoft Word, Pages for Mac, and Writer from Apache OpenOffice, or with a web-based program like Google Docs, you can create documents that include sophisticated formatting: for example, you can change text fonts (styles applied to text); apply effects such as bold, italics, and underlining; add shadows, background colors, and other design treatments to text and objects; and include tables, photos, drawings, and links to online content. You can also use templates (pre-designed documents with formatting and graphics already in place for you to fill in) to design web pages, newsletters, and more. A mail merge feature makes it easy to take a list of names and addresses and print personalized letters, envelopes, and labels. Figure 5.1 shows some of the word processing features and tools Microsoft Word offers.

Spreadsheet Software With **spreadsheet software**, such as Microsoft Excel, numbers rule. Using spreadsheet software, you can perform calculations that range from simple (adding, averaging, and multiplying) to complex (estimating standard deviations based on a range of numbers). In addition, spreadsheet software offers sophisticated charting capabilities. Formatting tools help you create polished looking documents such as budgets, invoices, schedules, attendance records, and purchase orders. With spreadsheet software such as Microsoft Excel, Numbers for Mac, and Calc (part of Apache OpenOffice), or with a web-based program like Google Sheets, you can also keep track of data such as your holiday card list and sort or search that list for specific names or other data. Figure 5.2 shows a typical spreadsheet using several key features.

FIGURE 5.1 **A Word Document**
Sophisticated word-processing tools allow you to create attractive documents.

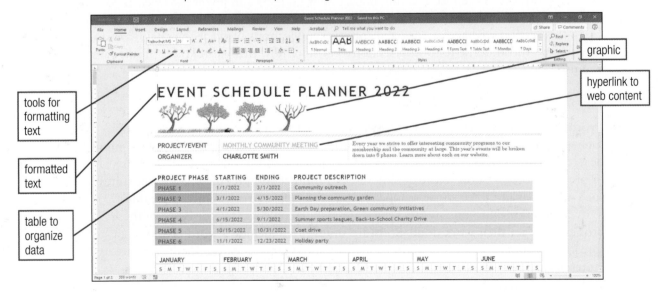

tools for formatting text

formatted text

table to organize data

graphic

hyperlink to web content

FIGURE 5.2 **An Excel Worksheet**

This worksheet in an Excel workbook tallies expenses automatically with formulas and includes an embedded chart to present the information visually for the reader.

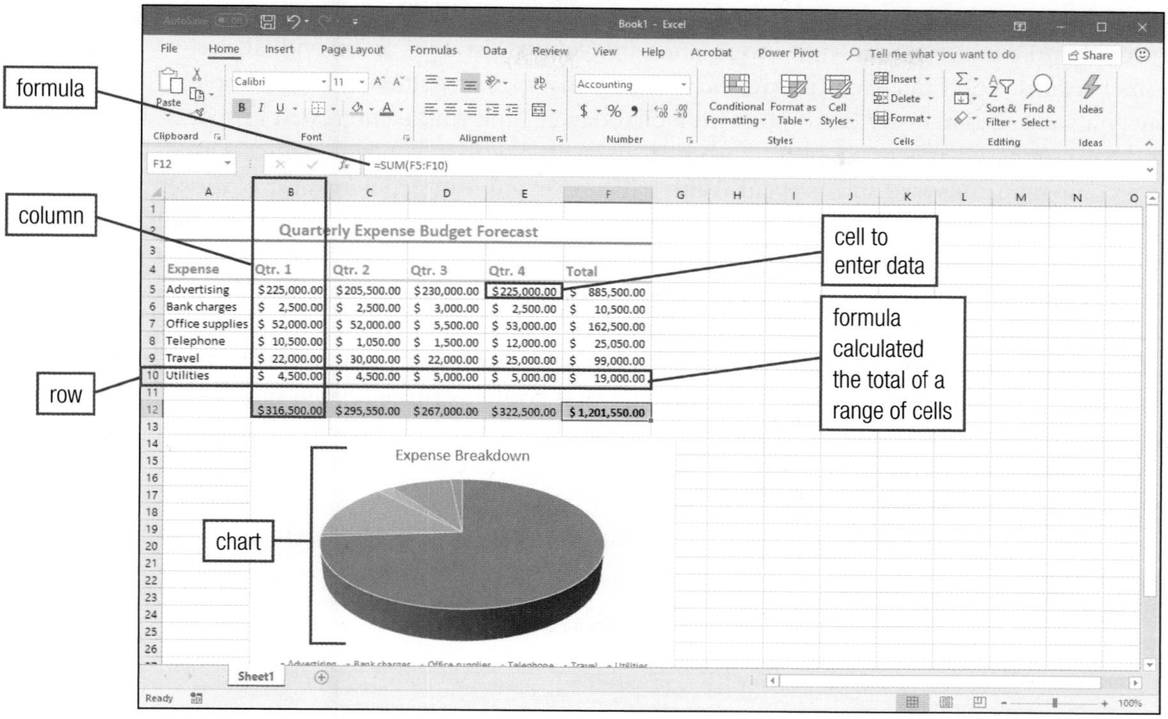

formula

column

row

cell to enter data

formula calculated the total of a range of cells

chart

Computers in Your Career

Spreadsheet software isn't just used by accountants. Managers use spreadsheets to compile and track department and project budgets. Statisticians and people working in marketing use the charting features of spreadsheets to map trends and summarize survey findings. Salespeople may use a spreadsheet to keep track of sales by customer or region. An administrative assistant in a small company may use the data functions to manage customer lists. The calculation, data manipulation, analysis, and charting features of spreadsheet software can be useful in many, many occupations.

Database Software **Database software**, such as Microsoft Access and Base (Apache OpenOffice), can manage large quantities of data. A **database management system (DBMS)** provides you with the ability to store data in a database and to update and retrieve that data securely. The software provides the functions for organizing the data into related **tables** and retrieving useful information from these tables. The data within each table is arranged into **fields** (one data item per field) and **records** (all the fields related to an item, such as a person or company). Each value that you enter in a field is called an *entry*. For example, imagine that you are a salesperson who wants to create a list of customers. Of course, you want to include fields to store the name, address, and company name for each person. However, you might also want each customer record to include the customer's birthday, his or her spouse's name, and his or her favorite hobby as well as a record of purchases in the past year. You can also set up fields to look up data in tables in the database. For example, you can use the zip code field to look up the name of a city that is stored in a different table. Using look-up fields saves time by eliminating the need to enter a city name over and over. Reducing redundant data also helps improve the accuracy of the database. Once that data is entered into a table you can view information in a tabular list or as an individual customer record. You can create **queries** that let you find specific data sets. For example, say you want to find every customer with a birthday in June who is interested in sports and has purchased at least $2,000 of products in the last year so you can invite them to a company-sponsored sports event. With a database, you can generate a list of those records quickly. Although hidden behind the scenes, all Access database queries are built using code that uses **Structured Query Language** (**SQL**, pronounced "sea-quel").

A **relational database** builds relationships among fields of data. For example, for a customer, the database might store first name, last name, home phone number, address, and what items each customer has purchased. Rather than try to put all the data into one table, you might build a database that contains three related tables: Customers, Zip Codes, and Purchases. In order to set up relationships in a database, every table has a unique primary key field. The **primary key field** ensures that each row or record of data in the table is unique.

Figure 5.3 shows some tools that Access uses to organize and manage data.

FIGURE 5.3 **A Microsoft Access Database**
Access offers several useful tools for managing data.

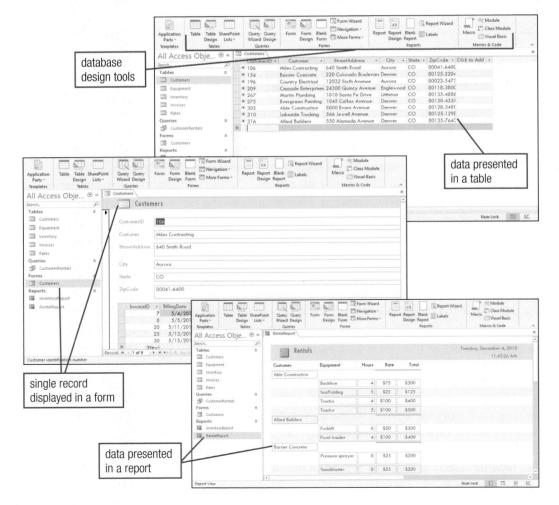

Presentation Software **Presentation software**, such as Microsoft
PowerPoint, KeyNote for Mac, Impress from Apache OpenOffice, or web-based programs like Google Slides, uses the concept of individual slides that collectively form
a slideshow. Slides may contain bulleted lists of key concepts, graphics, tables, animations, hyperlinks to web pages, and diagrams and charts. A slideshow can provide
visual support for a presenter's comments during a talk, run continuously on its own
(at a trade show booth, for example), or be browsed by an individual. Presentation
software, including some newer programs and online services such as Prezi and
PointDrive, helps users create attractive slides by allowing them to use templates containing background art and placeholders for objects such as titles and bulleted text.
Presentation programs also make it easy to add graphics to slides. Figure 5.4 shows a
presentation slide created in Microsoft PowerPoint and Figure 5.5 shows a different
slide created in KeyNote, which is designed for the Mac operating system.

Software to Keep Us Organized

Several software products help you organize your life by keeping track of the people
you deal with and your personal and work schedules. Many of these products are comprehensive, offering email, calendar, and contact management, like Google apps, and
Microsoft Outlook. Comprehensive personal organization software is usually available
in a web-based format and may also sync to a desktop or mobile app for ease of use.

You can use **calendar software** to schedule appointments or events and set up
reminders. The web-based Google Calendar, for example, lets you schedule events,
invite others to your appointments, and share your calendar with others. **Contact
management software** helps you store and manage information about the people

FIGURE 5.4 **A Microsoft PowerPoint Presentation**
Several available views in PowerPoint help you see your presentation as individual
slides, an outline of the content, as speaker's notes, or as a set of slides you can
easily reorganize.

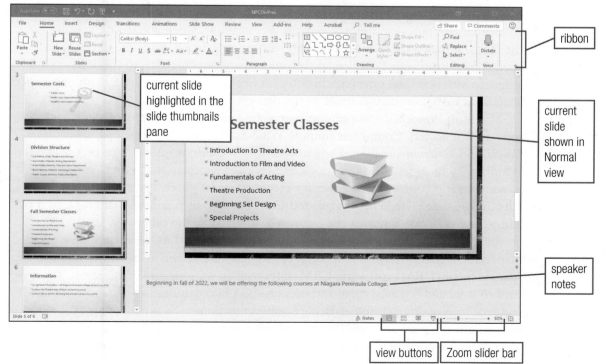

FIGURE 5.5 A KeyNote Presentation

Presentation software offers templates with design elements to create a polished look.

with whom you interact. You might keep track of information for clients, family members, or people in your book club, for example. Microsoft's Outlook.com is an online program that includes email, calendar, contact management, and storage features in one place. Many applications share data with other applications; for example, your contact app can provide addresses to be used in your email app. The Calendar and People features of Outlook.com are shown in Figure 5.6. Windows and the operating systems for most mobile devices include application software that has an address book function for storing contact phone numbers, email addresses, and the like.

Data sharing also allows you to share calendar event and contact information with your other computers and mobile devices. For example, when you leave the office for the day, you may want to make sure that all your appointments for the next day have been copied from your PC calendar to your smartphone calendar, so you'll have your schedule with you. You sync (synchronize) devices to make sure that data on one device is updated based on changes made to the data on another device.

Microsoft Windows is working to keep your calendar and contact management software connected. Computers running Windows 10 include three apps, called *Mail*, *People*, and *Calendar*, that work together and with Outlook online. The People app for Windows 10 includes an address book and provides access to social apps. You can use it to add and edit contacts, view posts on Facebook and Twitter, and communicate digitally using Skype. The People, Mail, and Calendar apps in Windows share data, so if you add a contact in one app, that contact will be available in the other two apps as well.

Still other applications and services, such as Salesforce and ACT!, expand on contact management and function. These suites of programs are known as **customer relationship management (CRM) software**. CRM suites contain software and online services that are used to store and organize client and sales prospect information, as well as to automate and synchronize other customer-facing business functions such as marketing, customer service, and technical support. This type of software can be useful for those who have to keep track of many customers or a lengthy sales or product implementation process.

FIGURE 5.6 **Outlook.com Calendar and People Features**

Use Calendar to schedule, edit, and view events, and use People to manage contacts.

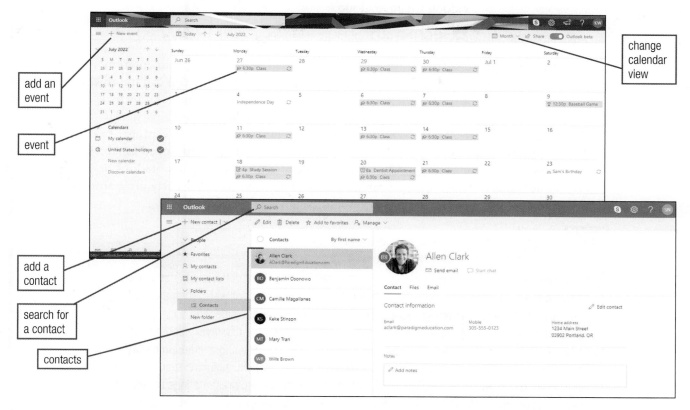

Graphics, Multimedia, and Web Pages

If you like to work with drawings, photos, or other kinds of images, you may have used **graphics software**, which is software that allows you to create, edit, or manipulate images. Though most types of productivity software, such as word processors and presentation software, include several graphics features, design professionals work with products that are much more feature-rich. These products include:

- **Desktop publishing (DTP) software**, which is used by design professionals to lay out pages for books, magazines, brochures, product packaging, and other print materials. Adobe InDesign, CorelDraw, and QuarkXPress are three of the most popular DTP products. These software applications allow a great deal of precision and control over the placement of objects and text on a page by using grids and columns, a large variety of fonts and text formatting tools, and the ability to insert, rearrange, and modify images.

- **Photo editing software**, which is used by design professionals to enhance photo quality or apply special effects such as blurring elements or feathering the edges of a photo. Photo editing software such as Adobe Photoshop and GIMP can also be used to edit content out of a picture or combine photo content into a more comprehensive composition.

Design professionals can choose from a variety of sophisticated software products.

 Adobe® Creative Cloud™

 Quark

TechSmith Snagit™

CorelDRAW X8

- **Screen capture software** is useful for those writing software documentation or books about software, programming, or other technology. You can use this software to capture all or a portion of the contents of the computer screen. You can then use the resulting image to demonstrate how to use a particular software feature. Recent Windows versions include a basic screen capture application called *Snipping Tool*. However, if you need more advanced capabilities, such as the ability to change the image resolution or add annotations or other edits to screen capture images, then an inexpensive third-party program such as Snagit or HyperSnap-DX might be a better choice.

Computers in Your Career

Even if you don't intend to become a professional designer, you are still likely to use design software in your personal or work life. Simpler desktop publishing software, such as Microsoft Publisher, is useful for designing flyers to advertise a yard or garage sale, find a lost pet, or create a home business brochure or business card. Shotcut is an entry-level video editing software program that is simple to learn and might be useful in creating a video promoting your company's newest product on YouTube, for example. Many jobs, including teaching and sales, require that you give presentations, and you can enhance the presentations by adding animations or music. No matter what your career path, consider exploring graphics and multimedia products to keep your work creative and current.

Still images aren't the only medium you can work with from your computer. **Multimedia software** enables you to work with media such as animations, audio, and video. The following are a few popular categories of multimedia software:

- **Animation software** enables you to animate objects and create interactive content (content the viewer can manipulate and control) that is sometimes combined with music or narration. Animations are like sophisticated cartoons that can be used in presentations and on the web to educate, advertise, or entertain. Popular animation software includes SmithMicro Software's Moho for 2-D animation and Autodesk's Maya for 3-D animation.
- **Audio software** such as Audacity or Sound Forge helps you work with music files and record and edit audio to create a **podcast** (an audio presentation that can be posted online) or other types of audio files to be shared with others.
- **Video editing software** such as Adobe Premiere Elements, Pinnacle Studio, and Shotcut are entry-level programs used to create and edit videos. Videos might include an audio track with voice or music, or a variety of special effects. Another product, Camtasia Studio, can record video of activity on your computer screen, which is useful for creating training videos for users. Professional level video editing programs can cost more than $1,000 and take significant time and training to master.

Another popular type of application software for personal and business use is **web authoring software**. Programs such as RapidWeaver for Mac developers and Adobe Dreamweaver provide advanced tools for creating web pages. Some of these products require knowledge of HTML (hypertext markup language) and advanced scripting technologies; others allow you to use a word processor–like interface to create text and add links, animations, and graphics to web pages. The current generation of web authoring programs is designed to adhere to or follow current standards for implementing HTML. These standards, established by the World Wide Web Consortium (W3C), define how to create pages that will display correctly in most web browsers. W3C standards include accessible pages to help ensure equal access to everyone. Rather than forcing you to work in code, most web authoring tools offer a **WYSIWYG (what you see is what you get)** interface. WYSIWYG (pronounced "wiz-e-wig") simply means that the way the contents look as you are designing the web page in the software is the way they will look in a browser.

Major multimedia program suites often include a web authoring program. For example, depending on which version you use, Adobe's Creative Cloud can include more than ten programs for working in print, interactive, audio, video, and web media.

Adobe Dreamweaver provides a way for web designers to create web pages without having to work with the underlying programming code.

Entertainment and Personal Use

Software isn't limited to the workplace. Huge industries are built around entertainment, gaming, and personal use software products.

Entertainment software is a category of software that includes games you play on your computer or game console. According to the Entertainment Software Association's 2017 Essential Facts about the Computer and Video Game Industry, 67% of US households own a device to play video games and 65% of US households have at least one person playing 3 or more hours of video games a week. The average age of gamers is 35 years, and women 18 years and older represent a larger portion of game-playing population than boys under age 18. All in, consumers spent $30.4 billion on video games in 2016.

But gaming software isn't just for fun; it's also used to improve physical fitness (as with FitBit) and to communicate about social causes. Social cause games are so popular that there is an annual conference for promoting and sharing them called the Games for Change Festival.

Software is also designed for educational purposes, available as both **edutainment** (software with both entertainment and educational value) and **web-based training** (training with some self-directed element accessed via the web). A **MOOC (massive open online course)** extends the concept of online courses to reach large numbers of individuals. MOOCs are typically open enrollment (anyone can enroll in the course), are free of charge, and have required assessments. Most MOOCs end when you have completed all the required readings, assignments, and assessments. Some colleges have agreed to accept MOOCs for credit and some MOOCs offer certificates that can be shared with employers to verify completion. Since its inception, this form of online education has been undergoing rapid changes in both definition and process.

🔒 Playing It Safe

If you download software such as games or media players, be sure you get them from a trusted source. If the source is questionable, you could be downloading viruses or spyware along with your game. Use antivirus and antispyware software such as McAfee or Norton, or a free product such as Windows Defender (built into Windows 10), AVG Antivirus Free, or Avast Free Antivirus. Run updates and scans often.

Games for Change promotes socially conscious games.

Hobby-related software covers almost any kind of interest, including genealogy, scrapbooking, sports, home design, and gardening. In some cases, what is a hobby to one person is a job to another, so you may end up using one or more of these products in your work as well.

If you want to organize your legal affairs or finances, many software products can help you. You can use specialized software to create your own legal documents such as wills, living trusts, or real estate leases. You can use tax software such as TurboTax to help you prepare your taxes. Personal financial packages such as Quicken allow you to track checking account activity, create and print checks, download checking account statements from your bank's website, reconcile your balance, and export information to tax software. In addition, you can track credit card accounts and investments.

Software Gets Professional

Some types of application software address specific business needs, such as bookkeeping, project management, or the processes used in industries such as hotel management or banking.

Financial Applications Financial software used by accounting professionals includes programs such as QuickBooks and Sage 50 Accounting. These accounting applications offer business-oriented features such as payroll, general ledger (the main accounting method for a business), invoicing, and reporting.

Business Processes Some software applications used in business address specific processes that are useful in a variety of industries. The following list describes a few such products:

- Project management software such as Microsoft Project or the open source program OpenProj is used in many settings, from construction to space flight, to plan the timing of tasks, resources, and costs. With sophisticated tools and algorithms to calculate schedules and costs, project management software can help track the many factors that can cause delays or cost overruns. Project management software can help a business save thousands of dollars, especially when it comes to large projects that can take years to complete.
- CAD/CAM stands for computer-aided design/computer-aided manufacturing. This category of software is used to create complex engineering drawings and geometric models and to provide specifications for the manufacture of products. CAD/CAM software is used extensively in industries such as automotive manufacturing, bridge and factory construction, and architecture. An example of CAD/CAM software is AutoCAD from AutoDesk.
- Purchasing software is used by people in any industry where a sophisticated purchasing process must be managed. That process begins when a purchasing agent seeks quotes for a particular item, generates a purchase order, tracks that the item has been received, and submits an invoice to accounting for payment. Purchasing software is also used to track inventory and place new orders so a company's supply of a vital part or material never runs out.

QuickBooks is an example of financial software used by accounting professionals. Financial software is used to keep track of information about a company's customers and the transactions that take place with those customers.

Microsoft Project is used in various industries to plan, track, and report on projects.

- Document management software and systems enable a business to better store and manage critical business information including electronic or scanned documents such as contracts, images, email messages, and other forms of data. These systems not only replace bulky physical storage such as filing cabinets, but also enable people to search for and retrieve documents without regard to physical or geographic location. Because these systems reduce paper usage, they are also more environmentally friendly than traditional filing systems.

Industry-Specific Software Many industries have specific software needs based on the day-to-day tasks their employees perform. For example, hotels, medical offices, travel businesses, real estate offices, and banks have different functions that require different specialized software packages.

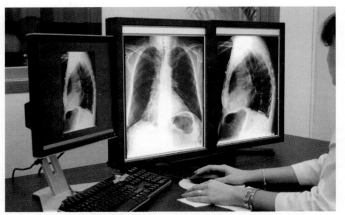

If you ran a large hotel, for example, you would need software such as Guest Tracker, which handles guest registrations, tracks charges such as room service and movies viewed, and generates a final bill. In the medical industry, specialized software handles insurance forms and billing, drug prescriptions for pharmacies, and electronic health record (EHR) management. Real estate professionals use property management software to keep track of rental properties and real estate sales software to analyze property values and track listings.

Many industries have their own unique computing needs that industry-specific software can accommodate.

Custom and Enterprise Software Larger companies (often referred to as *enterprises*) sometimes create their own software to handle the specific processes of their businesses. However, custom software is expensive to create and maintain, so many large companies adopt enterprise software such as SAP ERP, which addresses common business processes but is highly customizable. ERP stands for enterprise resource planning, which is a category of software that deals with standard business processes such as managing customer information (Customer Resource Management or CRM), product inventories (Supply Chain Management or SCM), and employee records and benefits systems (Human Resources Management or HRM).

Companies such as German-based SAP and US-based Oracle produce software modules that address these common business functions, building in best practices and standard forms and features. They then work with individual companies to customize those product features to the organization's specific procedures.

In recent years, enterprise software companies have broadened their offerings to include small and mid-sized business products, but depending on the degree of customization and training required, implementing an enterprise software product or suite of products at a company can be a very costly proposition. Still, when such products streamline and improve company procedures, the payoff in productivity and profit can be huge.

SAP enterprise products help track business activity and generate sophisticated analyses and reports.

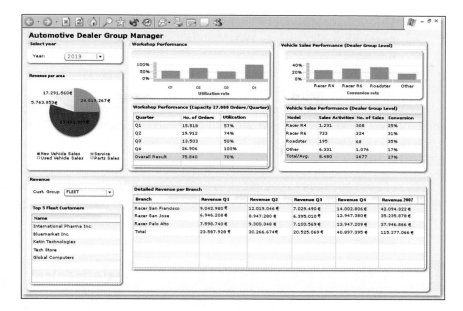

Mobile Applications

Today, everybody is mobile, so applications have come to your tablet or smartphone in a big way. In the past, mobile applications were typically not as full-featured as their non-mobile counterparts because of the limitations of memory and screen size, but today's smartphones and tablets feature much more power and storage than earlier hardware. As a result, many applications have been created for mobile devices, allowing you to perform a variety of activities while on the go. In addition, web designers have taken advantage of new technology standards to design sites that work well on many devices. So-called **responsive design** allows web pages to sense the device being used and reformat content for optimum display on all screen sizes.

For example, some productivity software, such as Word, Excel, and PowerPoint, is available in mobile versions; and various other mobile apps are designed to help us stay organized. Google has an entire suite of applications that includes productivity software, organizational software, and mapping. Other software for your mobile device can provide updates to the weather and news.

The list of the various types of mobile apps is as long or longer than it is for desktop apps. Here are some examples:

- Games
- Music
- Tools such as calculators, currency convertors, and music players

- Mobile banking
- Research tools (such as WikiMobile)
- Browsers
- Instant messaging
- GPS navigation applications

In 2018, the number of apps in the Apple App Store reached 2 million. Apps in the Google Play store exceeded 2.6 million in 2018. There were 178.1 billion downloads from mobile apps stores in 2017, and this number is estimated to grow to 258.2 billion by 2022.

> **Course Content**
> Take the Next Step Activities

5.3 Developing and Delivering Software

The way that software is developed and delivered has changed dramatically over the last several years. This is mainly due to the internet and the ability to download software applications quickly to your computer or use software hosted on the web. Other changes, such as the development of open source and shareware applications, have also had an impact on the cost of software for the consumer.

Developing Software

The **software development life cycle (SDLC)** has evolved over time. This procedure dictates the general flow of creating a new software product. As illustrated in Figure 5.7, the SDLC involves:

- Performing market research to ensure there is a need or demand for the product and then completing a business analysis to match the solution to the need.

> 66 Software undergoes beta testing shortly before it's released. Beta is Latin for 'still doesn't work.' 99
> —author unknown

- Creating a plan for implementing the software, which involves creating a budget and schedule for the project.
- Programming the software, which is the phase where software engineers and programmers create the actual program code.
- Testing the software, which involves having users work with the **alpha version** (the first stage of testing of the software) and then one or more **beta versions** (the second and subsequent stages of testing of a software product). Once all tests are complete and the product seems stable, a final **release to manufacturing (RTM) version** is produced.
- Deploying the software to the public, either by selling the product in a package or online, or by installing it on a company network or workstations, if it is a custom software product. Some software is entirely cloud-based, which means that users never install it but instead work with it online.
- Performing maintenance and bug fixes (a **bug** is a problem or malfunction in a program) to keep the product functioning optimally. During this phase, updates to the software may be released to resolve several maintenance or bug issues at one time. The update may be applied automatically or be offered as a service release that a user may choose to install or not.

FIGURE 5.7 The Software Development Life Cycle (SDLC)

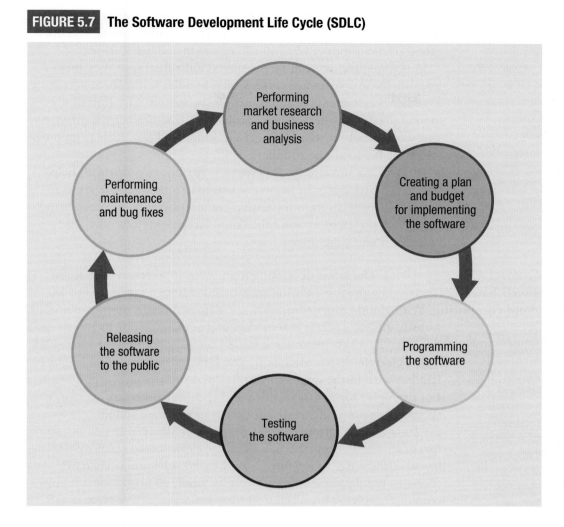

Today, software is often updated several times before a new version is released. These updates happen automatically over the internet.

How Software Is Delivered to You

There was a time when software only came in boxes that contained several floppy disks (thin magnetic discs in either flexible or rigid cases) and large, printed manuals. You might have been required to insert almost a dozen floppy disks into your computer to install a large program such as an operating system. Free technical support came with most products to help users install the software and troubleshoot problems when they arose. Later, software was released on CD and then DVD. Multiple floppy discs could be reduced to fewer CDs and eventually to one DVD.

Downloaded Software Today, software is typically downloaded from a website directly to your PC. A significant development in recent years is the ability to buy and download software online, from either the manufacturer or an online retailer. Before faster internet connections came along, transmitting huge application files was impractical, but today, with fast broadband connections, downloading a program can take just a few minutes. Downloading smaller apps to smartphones and tablets is even faster. Software can be free, cost a few dollars, or come with a hefty price tag. When you purchase software online you are often given a product key you can use to install the software.

Software products can be purchased individually, or they may be available in a **software suite**, which bundles several products together. Key benefits of suites include a common user interface among programs, the ability to easily share data among the products, and shared features that make certain functionalities, such as graphing, available to all the products.

Printed documentation has virtually disappeared. Instead, help files in the product itself or on the manufacturer's website are available, and free technical support has been replaced with tutorials and blogs on manufacturer and other sites, largely supported by user comments and suggestions.

Cloud Computing The latest development in software delivery and access is **web-based software**. The process of using web-based software is referred to as **cloud computing**. With cloud computing, software is hosted on an online provider's website and you access it over the internet using your browser—you don't have to have software installed on your computer to get work done. Google, IBM, Microsoft, and Oracle are names you might be familiar with in the list of companies that offer cloud computing. For example, Microsoft's Office 365 includes a word processor, a spreadsheet program, and a presentation program online.

Today, many software applications are available in the cloud, so users can often select if they wish to use cloud-based software or installed software. When using cloud-based software, users have access to the data and applications they need, all hosted on a remote computer server. This is referred to as **software as a service** (**SaaS**, pronounced "sass"). A provider licenses an application to customers to use as a service on demand. Office 365 from Microsoft is an example of SaaS. Office 365 offers online applications for personal, professional, and business needs for a subscription fee.

The cloud computing model has several advantages over installed software programs, including:

- There is no need to download a large application or take up room on your hard drive.
- Updates to the software product can be made frequently by the provider and are transparent (invisible) to the user.
- There is less danger of a conflict with another software product or driver on your computer.
- Applications can be accessed from any compatible device.
- Some cloud applications are free, while you may have to pay for the installed versions.

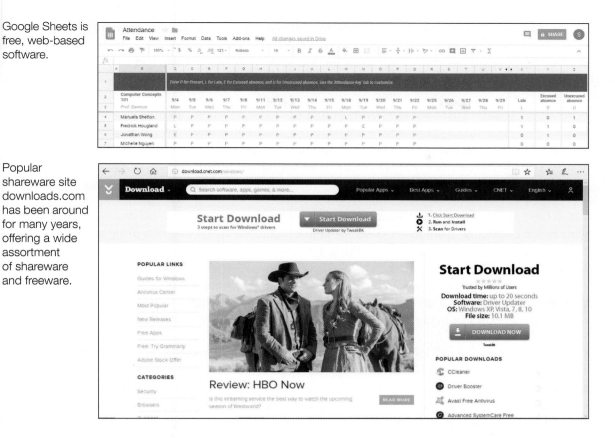

Google Sheets is free, web-based software.

Popular shareware site downloads.com has been around for many years, offering a wide assortment of shareware and freeware.

How We Pay for Software

Over the years, the cost of software has varied and continues to range from free to expensive, depending on the category of software.

Freeware and Shareware Generous programmers sometimes share software products for free (**freeware**) or for a small payment (**shareware**). Shareware lets you download a trial version for free for a limited time, but you have to pay to unlock all the features of the software. In recent years, the amount of free software available online has grown, and today you can probably find a free or shareware version of just about any type of application. Some of these free software products are quite sophisticated and feature-rich; others are a greatly simplified version of a similar software product you can purchase. Free products may also come with a hidden price: the company may capture and sell your information when you download the product, or they may use email and pop-up messages to attempt to sell you a more feature-rich product.

Open Source Software The open source movement makes **source code** (the programming code used to build the software) available to everyone in an effort to continue to build and improve the functionality of **open source software**. The open source movement has produced applications that are contributed to by individuals and are free to all. For example, WordPress is an open source web development system. Since there is no ownership of open source software, users with technical questions often have to rely on blogs and forums for answers. Service companies that charge a fee for technical support are sometimes formed to support open source software. The **GNU General Public License (GNU GPL)**, issued in 1989, specified policies about creating open source software, including that source code had to be made available to all users and developers.

Mobile Apps

There are thousands of mobile apps for use on smartphones and tablets. The apps can either be used for a fee or for free. Purchased apps are available for a fixed cost at the time of download. Although free apps may seem like a good deal at first, they do have a price, as they are often ad-supported or will encourage the user to make in-app purchases.

Licensed Software

When purchasing a software product, you will find a wide range of pricing, from lower costs for upgrading to a new version of a program, to higher, yet competitive, pricing for the full product at retail sites, auction sites, and manufacturers' sites.

Companies that wish to buy software for many employees enter into software licensing agreements that allow them to install a product on multiple workstations or on their network for individual computer users to access.

Cloud-Based Software

With the cloud computing model, vendors can charge a subscription fee. They host the application on their servers and provide **software on demand**, another term for SaaS. They can then disable access to the service when the user's contract expires. In some cases, use of the products is free, as with Google Docs, Microsoft Office Online, some Adobe products, and Amazon Cloud Player. Microsoft offers Office 365 by license with separate pay scales for students, single or multiple users, and corporate customers.

> **Course Content**
> Take the Next Step Activities
> Ethics and Technology Blog

5.4 Software Working Together

Today, many applications can use content created in other applications by importing or embedding the content. In addition, products within software suites often include shared features that provide functionalities such as diagramming or drawing.

Importing and Exporting

Many software applications allow you to **export** (send data to another application) and **import** (bring in data from another application). This works seamlessly within software suites such as Corel's WordPerfect Office suite and Microsoft Office. For example, PowerPoint includes an option to export a PowerPoint presentation to Word as handouts.

Other means of exporting and importing involve saving text in certain file types, such as **rich text format (RTF)**, which allows you to exchange data with many other applications. Saving as an RTF file reduces a document to a more rudimentary, commonly useable format that another software product is more likely to be able to open or import. Some software applications reduce the need for RTF files as they can directly open files that were created using a competing application.

Comma-separated-values (CSV) format, which separates each piece of data with a comma, is another widely used file type for importing and exporting. Using this format you might, for example, export all of your contacts from an Excel spreadsheet into a contact management application such as Sales Cloud from Salesforce.

Tablets are useful for handwriting notes when in a meeting or on the road.

You can determine a file's format by examining its **file extension**, those three or four letters following the period at the end of the file name, such as *xlsx* for an Excel file. Your computer operating system recognizes a file's format and what programs can open that file by checking these letters.

Converting Handwritten Notes and Speech to Text

Using **handwriting recognition software**, you can take handwritten input, such as notes you write on a screen with your finger or a stylus, and convert that content to a text file. Similarly, using Optical Character Recognition (OCR) software, you can scan a handwritten page and convert it to a text file. That text file can then be edited and saved using a program such as a word processor.

With **speech recognition software**, you speak into a microphone connected to or built into your computer. The software turns the sounds into text that you can then edit in a word processor. Speech recognition is built into Microsoft Windows, for example, and can be activated from within Microsoft Word.

Accuracy can be an issue with any OCR, handwriting recognition, or speech recognition software, as there are many variables in each person's handwriting, voice inflections, and accents. Generally speaking, handwriting recognition has advanced more rapidly than speech recognition software and can recognize nearly anyone's writing.

Embedding and Linking Content

Microsoft Office products use a technology called **object linking and embedding (OLE)** that allows content to be treated as objects that can be inserted into different software files, even if they were not created using that software. For example, you can place either a linked object or an embedded object, such as a graph from Excel, into a word processed document. A **linked object** maintains a link to the source; when the source is updated the linked item is also updated (if the source file is available). An **embedded object** resides in the file you've inserted it into along with information required to manage the object, but there is no connection to the object in the source file.

Using Shared Features

A **shared feature** is a small application that cannot run on its own but can be used with other software products. Shared features contain a particular functionality, such as the ability to build diagrams, and sometimes libraries of images that are useful to several kinds of software products. By allowing multiple products to draw on the feature, you save space on your computer and have a consistent procedure for getting that work done.

Some common shared features used by Microsoft Office products are:

- WordArt, which allows you to add shape, dimension, and color to text.
- SmartArt, which helps you diagram processes and work flow.
- Chart, which helps you build charts and graphs.
- Online Pictures, which is a library of artwork you can organize and insert into documents.
- Shapes, which is a collection of shapes you can draw in a document.

WordArt is a popular text enhancement feature that is shared by all Microsoft Office applications.

Course Content
Take the Next Step Activities

Course Content
The online course includes additional
training and assessment resources.

Summing Up

5.1 Software's Role in the World of Computing

Application software includes the software you use to get things done, from producing reports for work to creating art or playing games.

5.2 The Many Types of Application Software

Productivity software includes software that people typically use to get work done such as word processing, spreadsheet, database, or presentation software. This type of software is often compiled into suites of applications such as Microsoft Office, Apache OpenOffice, and Apple iWork. Productivity suites available in the cloud—notably Office 365 and Google Docs—also enable you to work wherever you have internet connectivity. Four major categories of productivity software are:

- **Word processor software**, which you can use to create documents that include sophisticated text formatting, tables, photos, drawings, and links to online content. You can also use templates to design web pages, newsletters and more.

- **Spreadsheet software**, which allows you to perform calculations that range from simple (adding, averaging, multiplying) to complex (estimates of standard deviations based on a range of numbers).

- **Database software**, which organizes data. Once that data is entered into relational tables, you can view information in a spreadsheet-like list or as individual records using forms. You can then create queries that let you find specific data sets.

- **Presentation software**, which uses the concept of individual slides that form a slideshow. Slides may contain bulleted lists of key concepts, graphics, tables, animations, hyperlinks to web pages, or diagrams and charts. A slideshow can support a presenter's comments during a talk, run continuously on its own (at a trade show booth, for example), or be browsed by an individual.
  Other categories of software include:

- **Calendar software** and **contact management software** that help you organize your time and professional or personal contacts, and **customer relationship management (CRM) software** that helps coordinate client-focused business functions.

- **Graphics software**, **web authoring software**, and **multimedia software** including **desktop publishing (DTP) software**, **photo editing software**, and **animation software** used by many professionals.

- **Entertainment software** such as games and special-interest programs related to genealogy or sports.

- Personal finance software for managing your spending and saving goals.

- Professional software to handle professional financial or business tasks. This includes software written for specific industry needs and software that packages business best practices in a customizable format.

- Software for mobile devices, which may be simpler versions of full-featured computer software. With the increasing power of smartphones and tablets, the variety and capabilities of mobile software continue to increase rapidly.

5.3 Developing and Delivering Software

Software delivery venues have changed dramatically over the last several years, mainly due to the internet and its ability to download software applications to your computer and allow you to use web-based software.

The process of developing software is described as the **software development life cycle (SDLC)**. Software typically comes out in new versions every few years. As it is being created by software engineers who write the source code, the new version goes through various development and testing phases—first an **alpha version** or very initial draft is released, and then one or more **beta versions**. When the software is in a final version, called the **release to manufacturing (RTM) version**, it is then duplicated or made available for download and sold to the public.

The latest development in software delivery is **web-based software**, generally referred to as **cloud computing**. With the cloud model, software is hosted on an online provider's website, and you access it over the internet using your browser. Software updates are usually done in the background as you work, over the internet. You typically access online software for free or by paying for a subscription.

In recent years, the amount of free software available online has grown, and today you can probably find a **freeware** or **shareware** version of just about any type of application. The **open source software** movement has produced many applications that enable the users to study the source code of programs, change and improve their designs, and distribute their updates.

Companies that wish to buy software for many employees enter into software licensing agreements that allow them to install a product on multiple workstations.

5.4 Software Working Together

Many software applications allow you to **import** and **export** data to other applications. This works seamlessly in software suites such as Corel WordPerfect Office and Microsoft Office. By saving files in certain formats, such as **rich text format (RTF)** and **comma-separated-values (CSV) format**, you can access data with many other applications.

Using **handwriting recognition software** or Optical Character Reading (OCR) software, you can take notes you handwrite and convert them to a text file that can then be used by word processors or other software. **Speech recognition software** turns words (sounds) spoken into a microphone into text that you can then edit.

Microsoft Office products and many other programs use a technology called **object linking and embedding (OLE)** to insert content created in one software application into different software files. Using linking, you can update the **linked object** and see those updates automatically reflected in the source file. With embedding, you insert the **embedded object** into a file and can edit it, but no link to the source content exists.

Shared features are small applications that cannot run on their own, but can be used from within other software products. Allowing multiple products to draw on the shared feature saves storage space on your computer and provides a consistent procedure for accomplishing certain tasks.

Terms to Know

5.1 Software's Role in the World of Computing

application software, 118

5.2 The Many Types of Application Software

productivity software, 119
word processor software, 120
spreadsheet software, 120
database software, 121
database management system (DBMS), 121
table, 121
field, 121
record, 121
query, 122
Structured Query Language (SQL), 122
relational database, 122
primary key field, 122
presentation software, 123
calendar software, 123
contact management software, 123
customer relationship management (CRM) software, 124

graphics software, 125
desktop publishing (DTP) software, 125
photo editing software, 125
screen capture software, 126
multimedia software, 126
animation software, 126
audio software, 126
podcast, 126
video editing software, 126
web authoring software, 127
WYSIWYG (what you see is what you get), 127
entertainment software, 127
edutainment, 127
web-based training, 127
MOOC, (massive open online course), 127
responsive design, 131

5.3 Developing and Delivering Software

software development life cycle (SDLC), 132
alpha version, 132
beta version, 132
release to manufacturing (RTM) version, 132
bug, 132
software suite, 134
web-based software, 134

cloud computing, 134
software as a service (SaaS), 134
freeware, 135
shareware, 135
source code, 135
open source software, 135
GNU General Public License (GNU GPL), 136
software on demand, 136

5.4 Software Working Together

export, 136
import, 136
rich text format (RTF), 136
comma-separated-values (CSV) format, 137
file extension, 137
handwriting recognition software, 137

speech recognition software, 137
object linking and embedding (OLE), 137
linked object, 137
embedded object, 137
shared feature, 138

Projects

Check with your instructor for the preferred method to submit completed work.

Spreadsheet Analysis

Project 5A

You have been assigned to create a spreadsheet that shows how your friends spend their days. Survey ten friends and ask them how much time they spend on the following activities: school, homework, sleeping, eating, socializing, and spending time with family. When you have finished collecting the information, set up the categories on a spreadsheet and enter your friends' responses. On the same spreadsheet, summarize the data. Submit the spreadsheet to your instructor.

Project 5B

Using the data from the spreadsheet you created in Project 5A, create a bar chart that displays your survey results. To complete this task, use the charting tools to select the type of bar graph you want to design. Format the chart and submit it to your instructor.

Using Web-Based Tools

Project 5C

Learning organizational and time management skills is important in today's busy society. As a student, you know firsthand the difficulties in balancing your classes, assignments, and possibly a job and family responsibilities. At times, you may forget assignments, appointments, or upcoming events. One tool that can help you stay organized is a Google Calendar. Create a school calendar to record all appointments, assignments, tests, and so on for the next two weeks. When you have finished scheduling your information, export the calendar and submit it to your instructor.

Project 5D TEAM

Your company's executive board is considering using Google tools such as Gmail and Google Drive to organize their communications and documents. The board members have asked your IT team to prepare a slide presentation that outlines the advantages of using these Google tools. Using Google Slides, create a collaborative presentation that discusses these Google tools and their applications. When complete, share the presentation with your class.

Graphics and Multimedia

Project 5E TEAM

Identify a software category that offers both a free product and a product you pay for, such as Apache OpenOffice and Microsoft Office. Create a table comparing the features of the two products, listing the entire feature set, and placing check marks next to the features each product includes. Consider both desktop and mobile versions of the product. Use your mobile device or a video recorder to create a video that presents the differences between the products and encourages the viewer to use the free software or purchase the commercial product. Submit the video to your instructor.

Project 5F TEAM

The public relations director of your school has asked the students in your Introduction to Computers class to create a flyer promoting the course. With your teammates, select a desktop publishing or word processing software application, such as Publisher or Word, to design your flyer. As you develop the content for the flyer, consider the basic information your flyer should provide: a general course description, the time and location of the class, the cost of the course, any course prerequisites, and the instructor's name. In addition to this general information, provide specific course objectives, learning tools, student projects, and activities that might serve as incentives for students to register for the course. To add visual interest, incorporate graphics by using tools such as WordArt.

Project 5G TEAM

Use the free Gmail photo sharing tool, Google Photos (https://ODW5.ParadigmEducation.com/GooglePhotos), to create a photo album scavenger hunt for your classmates. Have your team prepare a photo album of landmarks or sites in your local area. Ask each member of your team to contribute either an existing digital photo that they have or to go on location to take a new photo. Try to find photos that show enough detail to provide clues as to the photos' locations. Upload the individual photos to the team's album. When complete, share your team's album with the class. Ask other teams to post their responses as to the locations of the photos.

Word Processor and Presentation Software

Project 5H

Research how to create an effective slideshow presentation. As part of your research, read the article "Planning Is Key to Creating an Effective Presentation" at https://ODW5.ParadigmEducation.com/PresentationTips and click through the presentation "Designing Effective 'PowerPoint Presentations'" at https://ODW5.ParadigmEducation.com/EffectivePresentations.

Using the results of your research as well as your own personal experience creating or viewing slideshow presentations, make a list of ten recommendations or rules to follow when preparing a slideshow presentation. Submit your list of recommendations to your instructor.

Project 5I

You need to explain to a colleague how to change a font in Microsoft Word. In order to do this, you have decided to create a series of screen captures with callouts (descriptive labels). Use the software program Jing (available at https://ODW5.ParadigmEducation.com/Jing) or the Windows Snipping Tool to capture the screen images. After creating the images, insert the screen captures in a text document and add callouts.

Project 5J TEAM

You are a college student and a part-time employee of the IT department of a local business. Your manager recognizes your knowledge of current technological advances and would like you to share your insights with other members of the department. In particular, your manager would like you to explain cloud computing, the latest development in software delivery and access. Working with a team of colleagues, create a PowerPoint presentation explaining cloud computing. Post the presentation to Microsoft OneDrive or to a course website to share with your class and instructor.

Wiki Project—Collaborative Use of Software

Project 5K TEAM

Businesses today prefer to use software applications that allow their employees to collaborate on the creation of documents. To model that practice, your class will create a document for the class wiki on the advantages and disadvantages of buying a PC or a Mac. The document's content should include an analysis of features and applications available for each platform, the costs of components and any upgrades, and so on. Your instructor will divide you into three teams. Each team should post their content on the class wiki.

Project 5L TEAM

You or your team will be assigned to edit and verify the content posted by one team on the wiki site from Project 5K. If you add or edit any content, make sure you include a notation within the page that includes your name or team members' names and the date you edited the content—for example, "Edited by [student name or team members' names] on [date]." Keep a list of references that you used to verify your content changes. When you are finished with your verification, include a notation at the end of the entry—for example, "Verified by [student name or team members' names] on [date]."

Project 5M TEAM

You or your team will be assigned to format and add visuals to one team's document posted on the wiki site from Project 5L.

Learning about Application Software

Project 5N

As a help desk employee of a large insurance company, you often receive calls about how to use a variety of software applications. You have begun to keep a list of the most frequently asked questions and have decided to prepare a reference guide for employees that you will post to the company intranet. To help you with this task, go to Microsoft Office Online Help and find the answers to the following questions:

1. The data for my report is slightly confusing. I think that it would be better communicated in a table. How can I insert a table in Word to make my data more reader-friendly?
2. How do I create a chart for the current year's sales data?
3. Is there a way to create a formula using Excel to total the values of the range A1:A7?
4. I have to give a PowerPoint presentation to the Board of Directors, and I need to know how to export my Excel spreadsheet and import it to my PowerPoint presentation. Is this possible?
5. I want to run a query in my Access database showing all my customers in the US. How do I run an Access query?
6. I have two tables in Access, and I want to join the tables so that I can run a query with data from both tables. How do I join tables in Access?
7. All my colleagues use a variety of animations in their PowerPoint presentations.

I have an object that I would like to animate. How do I apply a custom animation to an object?

8. One of my colleagues uses the Outline View to create PowerPoint presentations. I am unable to locate how to create a presentation in Outline View. Are there specific steps that need to be followed?

9. I have two new customers that I would like to add to my contact list in Outlook. Is there a way to add contacts?

Project 5O TEAM

You are a member of the IT department in a company about to adopt an enterprise software solution. The vice president of information systems has assigned your group to create a PowerPoint presentation for training customer service employees on enterprise customer relationship management (CRM) software. Research enterprise CRM software, and create a slide presentation that describes the software and its benefits using sound and animation. Submit the presentation, along with a reference list of sources, to your instructor.

Preventing Software Failure

Project 5P

You are the project leader for a new software application. The coding is nearing completion and you are investigating ways to ensure you do not encounter a major software failure. You decide to research other software products that have encountered failure and consider preventative measures that your project team can use. Read the article "Top software failures of 2017" at https://ODW5.ParadigmEducation.com/SoftwareFailures. Choose one of the examples mentioned in the article. Consider the steps in the software development cycle and identify what you think went wrong with this particular product and what could have been done differently to prevent the failure from happening. Create a presentation that lists your recommended preventative measures along with the rationale for the recommendations. Be prepared to share the presentation with your class.

Class Conversations

Topic 5A   Is it OK to share software?

You have recently purchased the newest version of a cloud-based web authoring software application, which cost over $700. You are excited about using all the great features. Your friend also wants to use the same software but can't afford it. Your friend asks to log into the software using your credentials. You check the software terms of use and find that this is illegal. What would you do?

Topic 5B   Should I fix the photo to tell a better story?

Photo editing dates back to the 1860s. Manipulating photos can be done to misrepresent products, sensationalize something in the press, or cause somebody embarrassment. There have been several controversial photos that have been edited and placed in magazines or newspapers, such as a 1982 National Geographic cover photo of the pyramids in Egypt and a 2008 picture of vice presidential candidate Sarah Palin. Do you think that manipulating photos is ethical? Why or why not?

Topic 5C Can technology affect literacy?

Spell check is a feature available in many software applications. Most of you probably have grown up using spell check and grammar check. Are you dependent on spell check to proof your documents? Do you believe that features such as spell check and online calculators are causing our population to be overly dependent on them and unable to spell or calculate without those tools? What effect could this dependence have on the literacy of our society—or does it not matter how these things are checked, as long as the tools are available?

Chapter 6

Communications and Network Technologies
Connecting through Computers

What You'll Accomplish

When you finish this chapter, you'll be able to:

6.1 Differentiate between how networks are used to share data and computing resources in the workplace and at home.

6.2 List the components in a communications system and describe the types of signals and typical transmission speeds that travel over the system.

6.3 Recognize types of wired transmission media and list wireless transmission systems in use today.

6.4 Explain the role that network standards and protocols play in communications and give examples of commonly used wired and wireless networking standards.

6.5 Describe the three characteristics used to classify networks and give examples of typical network classifications.

6.6 Identify and differentiate among various networking devices and software that enable you to send and receive data.

6.7 State the reasons why network security is important and give an example of a security device.

6.8 Summarize trends that affect the future of networking.

Why Does It Matter ?

Your computers and mobile devices depend on a communications system to send and receive data over a network. Without such a system, you couldn't send emails and text messages, update your Facebook status, send a tweet, download music, or share a printer with other computers. Whether you're downloading a new app, sharing a file or a device with other computers in your home or office, or tapping into the resources on the worldwide network that is the internet, understanding how to use the power of networks to share with others can make your work easier and your life more satisfying.

 The online course includes additional training and assessment resources.

Computer networks allow people to connect to share work, devices, and information. Whether wired or wireless, networks use various technologies, hardware components, and software to make the connection.

Course Content

Take a Survey

Technology in Your Future Video

6.1 How Does the World Use Networking?

A **computer network** consists of two or more computing or other devices connected by a communications medium, such as a wireless signal or a cable. A computer network provides a way to connect with others and share files and resources such as printers or an internet connection.

> As network administrator I can take down the network with one keystroke. It's just like being a doctor but without getting gooky stuff on my paws.
>
> —Scott Adams, creator of *Dilbert*

In business settings, networks allow you to communicate with employees, suppliers, vendors, customers, and government agencies. Many companies have their own network, called an **intranet**, which is essentially a private internet within the company's corporate "walls." Some companies also offer an extension of their internal network, called an **extranet**, to suppliers and customers. For example, a supplier might be allowed to access inventory information on a company's internal network to make sure the company does not run short of a vital part for its manufacturing process.

In your home, networks are useful for streaming content from the internet and for sharing resources among members of your family. You may even have one or two smart appliances connected to the internet through your network. For example, via a home network, you might share one printer among three or four devices, stream music from your Spotify account to your smart speaker, and find out who just rang the doorbell with an app on your smartphone.

The internet is a global network made up of many networks linked together. If you consider all the applications, services, and tools the internet allows you to access, as you discovered in Chapter 2, you can begin to understand the power of networking and how it opens up a new world of sharing and functionality.

6.2 Exploring Communications Systems

A computer network is one kind of **communications system**. This system includes hardware to send and receive data, transmission and relay systems, common sets of standards so all the equipment can "talk" to each other, and communications software.

You use such a networked communications system whenever you send and receive text or email messages, pay a bill online, shop at an online store, send a document to a shared printer at work or at home, update a social network, and download music or videos. Figure 6.1 shows some of the common communications system components in use today.

The world of a computer network communications system is made up of:
- Transmission media upon which the data travels to and from its destination.
- A set of standards and **network protocols** (rules for how data is handled as it travels along a communications channel). Devices use these protocols to send and receive data to and from each other.
- Hardware and software to connect to a communications pathway from the sending and receiving ends.

The first step in understanding a communications system is to learn the basics about transmission signals and transmission speed when communicating over a network.

Types of Signals

There are two kinds of signals used in transmitting voices and other sounds over a computer network: analog and digital (Figure 6.2). An **analog signal** is formed by continuous sound waves that fluctuate from high to low. Your voice is transmitted as an analog signal over traditional telephone lines at a certain frequency. A **digital signal** uses a discrete signal that is either high or low. In computer terms, high represents the digital bit 1, and low represents the digital bit 0. These are the only two states for digital data.

Computers use a binary system of 1s and 0s, also called *digital signals* or *data* when processing and storing information. If you send data between computers using a medium that transmits data via an analog signal (such as older telephone and cable networks), the signal has to be transformed from digital to analog (modulated) and back again to digital (demodulated) to be understood by the computer on the receiving end. The piece of hardware that connects your computer to a transmission source such as

FIGURE 6.1 **Communications Systems**
Communications systems include computer hardware and communications software that allow computer users to exchange messages and other data around the house or around the world.

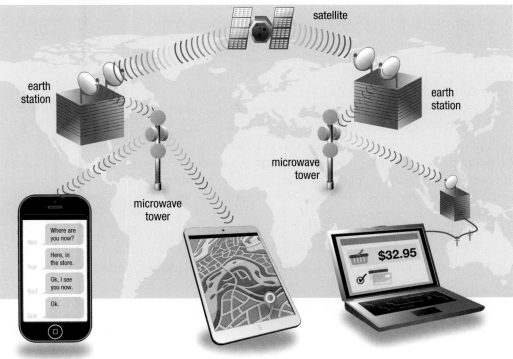

FIGURE 6.2 Analog and Digital Signals

A change in an analog wave represents a change in sound. Computers send and receive data using digital signals.

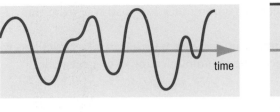

analog signal

digital signal

time

time

1   0   1   0   1   0   1

your telephone or cable television line to send and receive data is a **modem**. The word *modem* comes from a combination of the words *mo*dulate and *dem*odulate.

Today, most communications technologies use a digital signal, saving the trouble of converting transmissions. The television industry switched to digital signals in 2009, though some households still have an older television that has to use a converter box to convert digital transmissions back to analog. Computer networks use a digital signal for sending and receiving data over a network.

Transmission Speed

If you've ever been frustrated with how long it takes to download a file from a website, you're familiar with the fact that, in a communications system, data moves from one computer to another at different speeds. The speed of transmission is determined by a few key factors.

The first factor is the speed at which a signal can change from high to low, which is called **frequency**. A signal sent at a faster frequency provides faster transmission (Figure 6.3).

The other factor contributing to the speed of data transmission is bandwidth. On a computer network, the term **bandwidth** refers to the number of bits (pieces of data) per second that can be transmitted over a communications medium. Think of bandwidth as being like a highway. At rush hour, with the same amount of cars, a two-lane highway accommodates less traffic and everybody moves at a slower speed than on a four-lane highway, where much more traffic can travel at a faster speed. Table 6.1 lists communications bandwidth measurements.

Modems convert analog signals to digital signals and vice versa. Newer communications technologies use digital signals.

If you have plenty of bandwidth and your data is transmitted at a high frequency, you get faster transmission speeds. Any communications medium that is capable of carrying a large amount of data at a fast speed is known as **broadband**, such as cable, DSL, or fiber-optic.

Though transmission speeds at any moment in time may vary depending on network traffic and other factors, each of the common communications media has a typical speed. Table 6.2 lists typical transmission speeds. Providers continue to improve transmission speeds, and for some forms of connections, you can purchase plans offering much faster connections than those listed in Table 6.2. Some very high-powered connections being tested today provide transmission speeds of as much as 100 gigabits (one billion bits) per second, which allows you to download a high-definition movie in two seconds.

Most people quote connection speeds based on download speeds. Be aware that upload transmission speed is different from download transmission speed, and in many

FIGURE 6.3 **Faster and Slower Frequencies**
A signal that changes from high to low quickly travels faster.

short wavelength

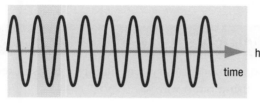

high-frequency signal

time

long wavelength

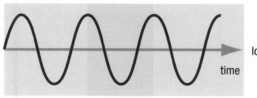

low-frequency signal

time

TABLE 6.1 **Bandwidth Measurements**

Term	Abbreviation	Meaning
1 **kilobit per second**	1 **Kbps**	1 thousand bits per second
1 **megabit per second**	1 **Mbps**	1 million bits per second
1 **gigabit per second**	1 **Gbps**	1 billion bits per second
1 **terabit per second**	1 **Tbps**	1 trillion bits per second
1 **petabit per second**	1 **Pbps**	1 quadrillion bits per second

TABLE 6.2 **Comparison of Network Connection Speeds**

Type of Connection	Typical Speed Range
dial-up	56 Kbps
satellite	12 to 25 Mbps
DSL	5 to 75 Mbps
cable	20 to 250 Mbps
fiber-optic	Up to 2 Gbps

cases the upload speed is significantly slower. Some companies even put restrictions on upload speeds.

Of course, everybody wants high-speed internet transmissions, but the higher the speed, the higher the cost. When you think about the type of connection you need for your computer, consider what you typically do when you're online. Do you send or receive many pictures or videos, which are typically very large files? Do you want to stream music or video, or play online games? In those instances, you need a broadband connection such as cable or DSL. On the other hand, if most of your transmissions are simply text messages or text documents, then a lower speed connection will probably work for you.

> **☁ Course Content**
> Take the Next Step Activity

6.3 Transmission Systems

The signals that are sent in a communications system might be transmitted via a wired medium (a cable) or wirelessly, using radio waves.

Wired Transmissions

At one time, all networks were wired. Long strings of wire connected each workstation to the network. Though they have lost popularity in the age of wireless, wired network models are still in use.

For example, in a business setting where many computers and printers remain in fixed locations, wired networks are still common. Many homes and businesses still use wires to connect network devices (such as routers) to a service provider, thereby providing a pathway for information transmitted over a network, called a **backbone**.

Wired transmissions send a signal through various media (Figure 6.4), including:

- **Twisted-pair cable**, which is used for an analog, wired telephone connection. This type of cable consists of two independently insulated wires wrapped around one another. Twisted-pair cable is used to transmit signals over short distances. It is used in many homes because signals can travel over the built-in telephone system. Cat5 is an example of twisted-pair cable used in early wired networks that is now considered obsolete. Cat5e (called *Cat5 enhanced*) can transmit data at speeds 10 times faster than Cat5. Newer wired networks may use Cat6, Cat6a, or even Cat7 twisted-pair cables, which can transmit data at lightning fast speeds ranging from 10 Gbps to 100 Gbps (Cat7).
- **Coaxial cable**, which is the same cable used to transmit cable television signals over an insulated wire at a fast speed—in this case, millions of bits per second.
- **Fiber-optic cable**, which uses a protected string of glass that transmits beams of light. Fiber-optic transmission is fast, transmitting billions of bits per second. Internet service providers are continually expanding their fiber-optic networks. Many now offer fiber-optic speed to home subscribers—an option originally only available to businesses.

Many businesses lease T-lines (such as T1 or T3), developed by long distance telephone companies, to carry multiple types of signals (voice and data) at very fast speeds over fiber-optic or older copper-wired lines.

FIGURE 6.4

Three Types of Cables

Wired transmissions send signals over coaxial, twisted-pair, or fiber-optic cables.

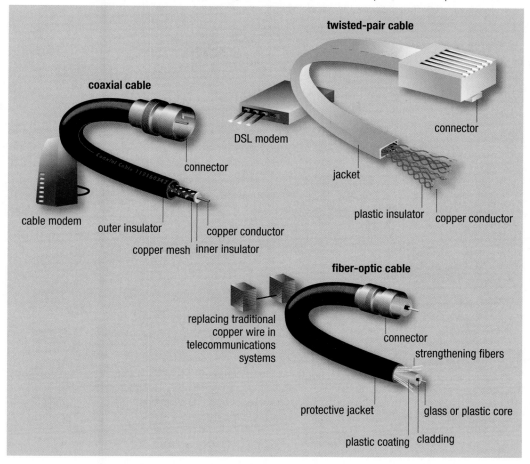

Wireless Transmissions

There are several wireless transmission systems in use today, including cellular, micro-waves, and satellite. All use radio waves to transmit data, but each system handles these transmissions in different ways, and the systems vary in signal strength and frequency.

A **cellular network**, like those used by your cell phone, transmits signals, called **cellular transmissions**, by using cell towers (Figure 6.5). Each cell tower has its own range (or cell) of coverage. Cellular systems are used to transmit both voice and data in every direction. There are several generations of cellular transmissions, from the first generation (1G) that was used for analog transmissions to the fourth generation (4G) which transmits digital data at speeds up to 100 Mbps. (Speed performance varies by carrier.) In 2018, carriers began to deploy 5G networks, bring-ing potential speeds to the wireless world that are 30 to 50 times faster than 4G (according to Verizon's 5G trials). By the end of the decade, 5G smartphones are expected to be available in many areas in the US and Canada.

A **microwave** is a high-frequency radio signal that is directed from one microwave station tower to another. Because the signal cannot bend around obstacles, the towers have to be positioned in line of sight of each other, as shown in Figure 6.6. Microwave transmission might be used, for example, to send a signal from towers located on top of several buildings in a school or company campus. If it's not possible to place towers

FIGURE 6.5 **Cellular Technology**
A smartphone sends a signal that travels along a series of cell towers until the data reaches the intended recipient.

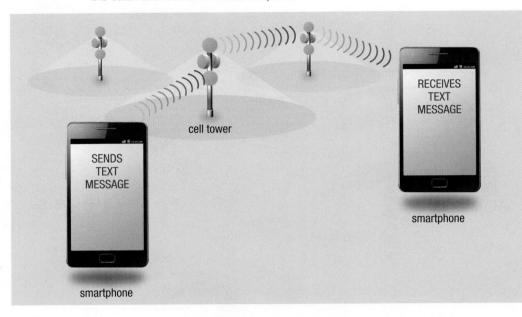

FIGURE 6.6 **Microwave Technology**
Radio signals travel through the atmosphere from one microwave tower to another. These towers have to be relatively close to each other because radio signals can't bend around the earth or around objects such as buildings. Alternatively, they can be bounced off a satellite.

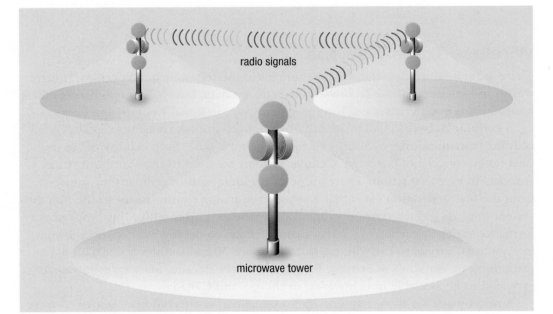

Our Digital World

To avoid physical obstructions between them, some microwave towers send signals up to a satellite to be bounced down to the target tower.

within sight of each other, signals can be sent upward where there are no obstructions. The signal is bounced off the atmosphere or a satellite and sent back down to another microwave tower.

Satellite communication uses space-based equipment, and is typically used for longer range transmissions, for international communications, and for connectivity in rural areas where cellular or microwave towers aren't available. A satellite receives microwave signals from an earth-based station and then broadcasts the signals back to another earth-based station or dish receiver (Figure 6.7). Video conferences and air navigation control are typical uses of satellite communication.

On March 1, 2018, NASA launched their second next-generation weather satellite for the National Oceanic and Atmospheric Administration (NOAA). These satellites provide near real-time weather data faster and more accurately than older satellites, improving NOAA's ability to track storm systems, lightning, wildfires, and other weather events that affect the US and its territories. Tracking and Data Relay Satellites (TDRS) have also been used for several years for services such as allowing doctors in urban areas to monitor surgeries or provide medical assistance to people in remote areas of the world.

FIGURE 6.7 **Satellite Communications**
Communications satellites are solar powered. They use transponders to receive signals from the ground and then retransmit them to another location on the ground.

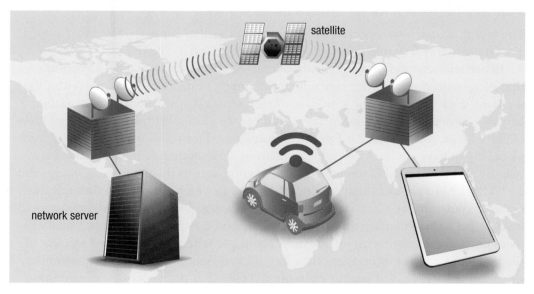

Course Content
Take the Next Step Activity

6.4 Communications Standards and Protocols

Browse through any computer store, online or off, and you'll find that there are many models of computers and other internet-connected devices made by many different manufacturers. These different hardware models have to have a way to communicate with one another. For example, you may want to send a text message from your Nokia smartphone to your friend who uses a Samsung Galaxy. Or, you may attach a document to an email message and send it from your HP laptop to a customer who uses a Mac. You may even want to send data or a picture from your smartphone to your computer.

To allow all these different devices to talk to each other, the computer industry has developed **standards** that address issues of compatibility among these devices. Organizations such as the **American National Standards Institute (ANSI)** and the **Institute of Electrical and Electronics Engineers (IEEE)** (pronounced "eye-triple-e") develop and approve these standards, which specify how computers access transmission media, speeds used on networks, the design of networking hardware such as cables, and so on.

A standard that specifies how two devices can communicate is called a protocol. A protocol provides rules such as how data should be formatted and coded for transmission.

Hardware manufacturers design their devices to meet network standards that relate to the types of tasks their devices handle. Data sent from one device to another travels across a wired system, a wireless system, or a combination of both wired and wireless using a variety of standards.

Wired Networking Standards

The way that data is transported within a network relates to the standard that is used. Two standards you should know about are Ethernet and TCP/IP.

Ethernet is a common wired media standard. The Ethernet standard specifies that there is no central device controlling the timing of data transmission. With this standard, each device tries to send data when it senses that the network is available. Ethernet networks are fast, inexpensive, and easy to install. With Ethernet speed capable of 1 Gbps, it is the standard most commonly used for wired networks in the workplace. If you have a high-speed cable modem or DSL modem at home, chances are you use an Ethernet cable to connect a wired computer to the networking equipment.

TCP/IP is a protocol that specifies the order in which data is sent through the network. With the TCP/IP standard, instructions are given to divide data into small units called *packets* that are passed along the network. A **packet** is simply a smaller piece of data combined with information about the sender and intended receiver of the data. The process of breaking data into packets, sending, and then reassembling the original data is called **packet switching**. TCP/IP is the network standard upon which internet communications are based.

Wireless Networking Standards

Getting rid of wires gives people the freedom to move around and make connections on the go. This process is important whether you're part of a mobile workforce, or you just like to use Twitter or email from your local coffee shop. Improvements in wireless technologies continue to enable mobile users to increase their access to different types of data, but just how do they work?

Wireless transmissions can use a variety of networking standards, including Wi-Fi, Long Term Evolution (LTE), Bluetooth, and Radio Frequency Identification (RFID).

Wi-Fi refers to a network that is based on the **802.11 standard** in its various versions, such as Wi-Fi 4, Wi-Fi 5, and Wi-Fi 6. Wi-Fi is a popular choice for setting up a wireless home network. This standard tells wireless devices how to connect with each other using a series of access points and radio frequencies to transmit data. Table 6.3 lists the differences among various Wi-Fi standards as well as the original names for standards released after 2009, which were renamed in 2018 by the Wi-Fi Alliance to simplify equipment identification.

Computers and portable computing equipment come equipped with built-in Wi-Fi capability. Many public spaces, such as libraries, airports, coffee shops, restaurants, and malls, are equipped with Wi-Fi access points so you can connect with the internet from your laptop, tablet, or smartphone. A location, such as a coffee shop, that makes Wi-Fi access available is called a **hotspot**. Operating systems such as Windows and macOS are capable of automatically detecting hotspots and connecting to them.

Long Term Evolution (LTE) is touted as the true 4G network. LTE is offered under a variety of banners by carriers such as LTE, LTE-Advanced, and LTE-A to name a few. LTE networks encompass a set of standards that involves changes to the wireless infrastructure to improve data speeds.

> **🔒 Playing It Safe**
>
> Networks can be secured or unsecured. When you're logged on to a free wireless hotspot such as the one in a local coffee house, your connection is not secure because public networks typically do not have security features turned on. Never perform financial transactions, such as checking your bank account balance or providing your credit card or checking account information to an online store, when using a hotspot. Always secure your home wireless network with a strong access password.

TABLE 6.3 **Differences Among Various Wi-Fi Standards**

Standard	Released	Frequency	Bandwidth	Highest Data-Rate
802.11 (Legacy)	1997	2.4 GHz	20 MHz	2 Mbps
802.11b	1999	2.4 GHz	20 MHz	11 Mbps
802.11a	1999	5 GHz	20 MHz	54 Mbps
802.11g	2003	2.4 GHz	20 MHz	54 Mbps
Wi-Fi 4 (802.11n)	2009	2.4 GHz, 5 GHz	20 MHz, 40 MHz	450 Mbps
Wi-Fi 5 (802.11ac)	2013	5 GHz only	20 MHz, 40 MHz, 80 MHz & 160 MHz (opt)	1.7 Gbps
Wi-Fi 6 (802.11ax)	2019	2.4 GHz, 5 GHz, and other bands as they become available	20 MHz, 40 MHz, 80 MHz, 80+80 MHz & 160 MHz	14 Gbps

Source: Adapted from *The Future of Next-Generation Tablets and Smartphones* by Partha Murali, Redpine Signals, Inc. on the Embedded Intel Solutions site at http://www.embeddedintel.com.

Bluetooth headphones are popular with smartphone users who want a wireless operation.

Bluetooth is a network protocol that offers short-range connectivity (3 to 300 feet, depending on a device's power class) via radio waves between devices such as your smartphone and car. Bluetooth-enabled devices can communicate directly with each other. Most cars now come with Bluetooth connectivity integrated into their systems for hands-free phone operation. This is partially the result of more and more states and provinces enacting laws against talking on a handheld phone while driving. A variety of devices today contain a Bluetooth chip, such as laptops, smartphones, mice, keyboards, and even digital cameras.

Bluetooth LE (BLE), which stands for Bluetooth low energy (also referred to as **Bluetooth Smart**), is a Bluetooth standard that uses much less power to communicate within the same range as what is now called "Classic" Bluetooth. **Bluetooth 5** devices that incorporate the newest Bluetooth standard became available in 2017. With Bluetooth 5, you can play audio on two connected devices with data transfer speeds up to 2 Mbps. Bluetooth 5 also uses an expanded connectivity range of up to 800 feet.

When two or more wireless devices are connected via Bluetooth, a **piconet** is formed. A piconet is also sometimes called a *personal area network (PAN)*. Piconets are temporary networks that send and receive data to and from up to eight Bluetooth connected devices operating on the same channel.

Smartphone users may be able to share the internet connection of their device via a cable, Bluetooth, or Wi-Fi with another device such as a tablet or laptop. This process is known as *tethering*, and it is useful when access to the internet via other means is not available for your laptop or tablet. In this case, your mobile phone basically becomes a modem that can service several other computing devices.

Radio Frequency Identification (RFID) is a wireless technology used primarily to track and identify inventory or other items via radio signals. An RFID tag contains a transponder (the part that sends the signal) which is read by a transceiver or RFID reader. As the tag is read, inventory data is automatically updated on the network. The manufacturing and retail worlds track products from production line to end consumer using RFID tags. This technology is also becoming more widely used as industries leverage it to track patients, pets, wildlife, access to parking gates, and toll payments, to name a few applications.

Have you paid for a coffee lately by tapping or hovering a debit card, credit card, or your smartphone near a payment machine? **Near Field Communication (NFC)** is the technology that makes such contactless payments. NFC works with apps including Apple Pay and Android Pay and uses an RFID frequency for close-range communications. When a customer holds or taps a card within a few inches of an NFC-enabled reader, the device and card exchange encrypted data in less than a second. Besides paying for purchases, NFC can also be used to unlock doors and pay transit fares.

Wireless Application Protocol (WAP) specifies how mobile devices such as smartphones display online information, including maps and email. Devices using WAP to display web content have to contain a microbrowser, a less robust version of a browser such as Google Chrome.

6.5 Network Classifications

Networks have three important characteristics that we'll look at in this section:
- The size of the geographic area in which the network functions.
- The role of computers in the network and how data is shared and stored on the network.
- How devices in a network are physically arranged and connected to each other.

Types of Networks

Networks can be set up to work in your house, your city, or around the world. There are three main types of networks, categorized by the area they cover and identified by the catchy acronyms LAN, MAN, and WAN.

Local Area Networks A **local area network (LAN)** is a network where connected devices are located within the same room or building, or in a few nearby buildings (usually not more than a hundred or so feet apart) (Figure 6.8). One computer is designated as the server. In computing terms, a **server** is any combination

FIGURE 6.8 **A Local Area Network (LAN)**
A switch or hub connects individual devices on a network.

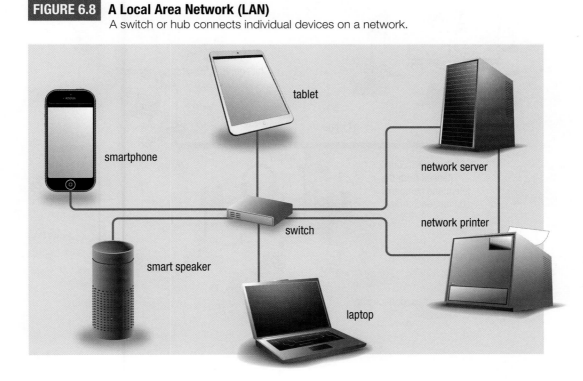

tablet

smartphone

network server

network printer

switch

smart speaker

laptop

of hardware and software that provides a service, such as storing data, to your computer. The server in a LAN houses the networking software that coordinates the data exchange among the devices. Shared files are generally kept on that server. A LAN can include a separate print server that manages printing tasks sent from multiple computers to a group printer. Small and home businesses are often networked in a LAN, which can be wired or wireless. A **wireless LAN** is called a **WLAN**.

Metropolitan Area Networks A **metropolitan area network (MAN)** is a type of network that connects networks within a city, university, or other populous area to a larger high-speed network. MANs are typically made up of several LANs that are managed by a network provider. If you connect to a network through your phone or cable company, then you probably connect through a MAN.

Wide Area Networks A **wide area network (WAN)** services even larger geographic areas (Figure 6.9). WANs are used to share data between networks around the world. The internet is, in essence, a giant WAN. WANs might use leased T1 or T3 lines, satellite connections, radio waves, or a combination of communications media.

Network Architecture

In the noncomputer world, architecture relates to the design of a building—where doors go, how walls connect to each other, and so on. **Network architecture** relates to how computers in a network share resources. The two major architectures are client/server and peer-to-peer (illustrated in Figures 6.10 and 6.11, respectively).

FIGURE 6.9   **A Wide Area Network (WAN)**
Three branches of a company located in different cities share resources through a WAN; a router directs network traffic to each city.

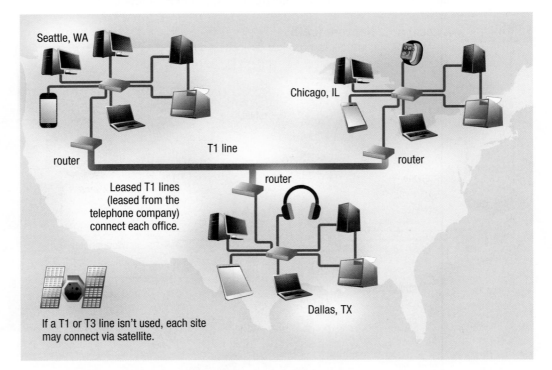

FIGURE 6.10 Client/Server Architecture

A client sends information to or requests a service from a server. The server sends the information on to another client located on that same network or processes the service request.

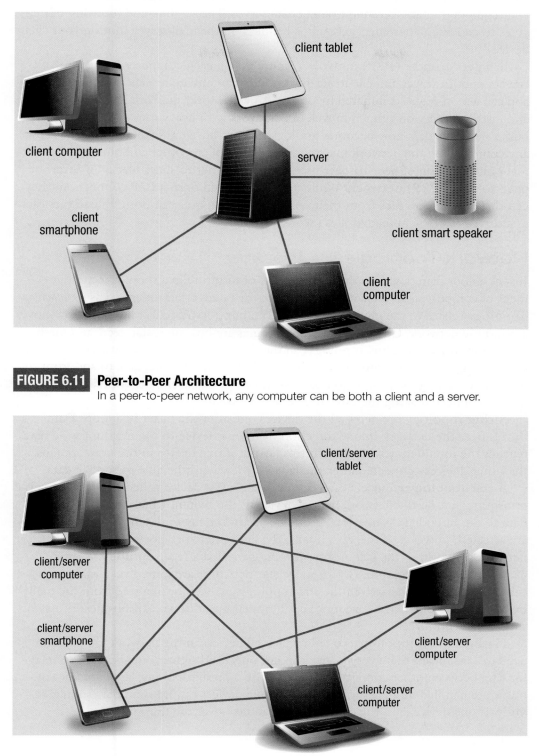

FIGURE 6.11 Peer-to-Peer Architecture

In a peer-to-peer network, any computer can be both a client and a server.

In a **client/server network** (Figure 6.10), the server stores programs and files that any connected device (called a **client**) can access. Client/server is considered to be a **distributed application architecture** because it distributes tasks between client computers and server computers. For example, on the internet, a typical client, such as your computer, makes a request of an email server. The sending and receiving of messages to and from your email account is handled on the mail server (not on your client device).

In a **peer-to-peer (P2P) network** (Figure 6.11), each computer in the network can act as both server and client. Each computer can initiate requests of other clients and can act on requests initiated by others. A peer-to-peer network is simple to design and install, but it won't function well in a network with heavy user demand. In a busier network, data flow becomes gridlocked like a busy traffic intersection, because any computer on the network may request services of any other computer.

A modification of peer-to-peer used on the internet to share files is an **internet peer-to-peer (P2P) network**. While users are logged onto a P2P connection, they can share and access files from their computers. Each user can both upload files (act as a server) or view uploaded files (act as a client).

Network Topologies

Just as a floor plan describes the arrangement of furniture in a room, the arrangement of computers, servers, and other devices in a network is described as its *topology*. **Topology** relates to the physical layout of data but doesn't reflect how the data moves around the network. Commonly used topologies are bus, ring, star, tree, and mesh (see Figure 6.12). In a **bus topology**, all the devices are connected by a single cable, like a string of holiday lights. The cable, called the *bus*, transmits data to and from devices by reading the destination address. If one device in the network fails, the other devices connected along the bus cable remain working.

A **ring topology** connects all the devices one after the other in a loop. Data travels over the network in one direction from one device to the next until the data reaches the intended destination. In this layout, if one device in the network fails, the other devices connected after the failed device won't be able to receive data.

A **star topology** connects each device to a central device called a *switch*. Data is transmitted along the network from a device to the switch, which then passes the data to the intended destination. If one device in the network fails, no other devices are affected; however, if the switch fails, the entire network goes down. For a larger network, a hybrid of the star and bus topology, called a **tree topology**, is often used. In this configuration, two or more star networks are joined via a main bus cable. Think of each switch in the star topology as the root of a tree from the main bus cable. These networks are more easily scaled up or down as a business grows or shrinks its workforce.

Finally, a **mesh topology** is one in which all devices in the network are interconnected. This means that every device can not only send signals but also act as a relay for data between other connected devices. With a mesh topology, large volumes of traffic are handled more easily because multiple devices can transmit data simultaneously. If one device fails, the other devices can take over the transmissions.

FIGURE 6.12 **Common Network Topologies**
The bus, ring, star, mesh, and tree topologies are common arrangements for connecting computers, printers, and network devices.

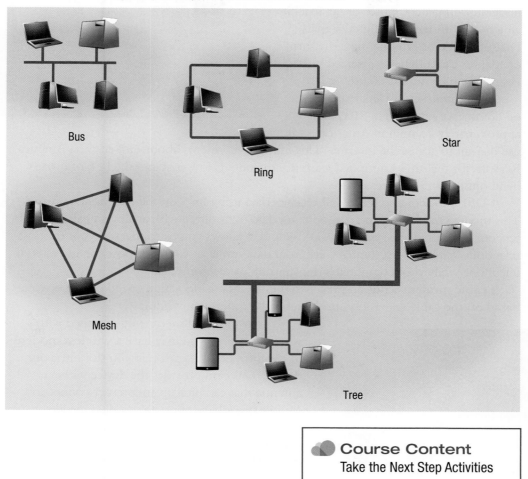

Bus

Ring

Star

Mesh

Tree

Course Content
Take the Next Step Activities

6.6 Networking Devices and Software

It's useful to understand the physical elements of a network in case you set up one yourself someday. Setting up a network involves several types of hardware and may involve a network operating system.

Network Devices

Each device connected to the network is called a **node**. Networking devices include two kinds of hardware: one that facilitates the exchange of data from your computer to a transmission medium such as a cable connection, and one that enables various devices on a network to communicate with each other.

Modems to Send and Receive Signals As you learned earlier, a modem is the piece of hardware that sends and receives data to and from a transmission source such as your telephone line or cable television connection.

A **dial-up modem** works with phone transmissions. These devices change or manipulate an analog signal so that it can be understood by a computer or fax machine, which only understand digital signals. Dial-up modems typically take the form of adapter cards that are inserted into your computer motherboard.

A modem allows your computer to send and receive digital data via a broadband connection. Newer hardware combines a modem, router, and wireless access point all in one piece of hardware.

The United States Department of Commerce reported in 2017 that approximately 22 million Americans still use dial-up to access the internet. This number represents about 35% of rural America. Reasons for dial-up's continued existence include its low cost and the lack of broadband options in rural areas.

A **DSL modem** also sends and receives data using lines on the telephone network. This modem modulates and demodulates data transmitted with analog signals, and filters out incoming voice signals. A DSL modem allows you to connect to your existing telephone system and separates voice from data traffic so you don't lose the use of your telephone while your computer is transmitting or receiving data.

A **cable modem** sends and receives digital data using a high-speed cable network based on the cable television infrastructure found in many homes.

A mobile broadband stick is a USB device that acts as a wireless modem to give your computer access to the internet.

If your mobile device is not equipped with a wireless adapter, you may have a **wireless modem** in the form of a PC card that you slot into your device. This card provides the device with an antenna that can pick up an internet connection. Mobile broadband sticks (also called *dongles*) have become popular for mobile users who want broadband flexibility while on the go. A **mobile broadband stick** is a USB device that acts as a modem to give your computer access to the internet. These wireless modems can be moved easily between devices. Powered by the computer itself, a mobile broadband stick does not need to be recharged.

Hardware That Provides Access to a Network A **network adapter** is a device that provides your computer with the ability to connect to a network. A **network interface card (NIC)** (pronounced "nick") is one kind of adapter card. The NIC processes the transmission and receipt of data to and from the communications system.

In most recent computers, NICs take the form of a circuit board built into the motherboard. NICs enable a client computer on a LAN to connect to a network by managing the transmission of data and instructions sent by the server. At home, you plug one end of a cable into your desktop or laptop NIC card and the other into a DSL or cable modem.

A **wireless interface card** functions in the same way as a NIC except that it connects wirelessly.

> "Describing the internet as the Network of Networks is like calling the Space Shuttle, a thing that flies"
>
> —John Lester

Hardware That Connects Devices to Each Other on a Network

There are several devices that help other devices on a network to communicate with each other.

A **wireless access point** is a device that contains a high-quality antenna (Figure 6.13). This antenna allows computers and mobile devices to transmit data to each other or to exchange data with a wired network.

A **router** is a device that allows you to connect two or more networks in either a wired or wireless connection (in which case it is referred to as a **wireless router**). On a home network, for example, a router allows you to connect multiple devices to the internet using one high-speed connection. Routers used to connect business networks have many ports and are faster and more sophisticated than the routers you use in a home network. A mobile hotspot combines cellular network access with a wireless router, enabling multiple computers or devices to have wireless internet access. Older models worked on 3G networks, while the newer models support 4G LTE mobile broadband. Expect to replace a router when 5G service becomes available to take full advantage of the faster speed and quality in 5G transmissions.

A **repeater** is an electronic device that takes a signal and retransmits it at a higher power level to boost the transmission strength. A repeater can also transmit a signal to move past an obstruction, so that the signal can be sent further without degrading, or losing quality.

A **switch** joins several nodes together to coordinate message traffic in one LAN by checking the data in packets it receives and delivering each packet to the correct destination.

Gateways and bridges are devices that help separate networks to communicate with each other. A **gateway** is used when the two networks use different topologies, and a **bridge** is used to connect two networks using the same topology.

A router (top left), switch (top right), repeater (bottom left), and bridge (bottom right) are devices that networks may use to connect and send data.

FIGURE 6.13
Wireless Access Point in a Network

A wireless access point sends the data to computers or other devices equipped with wireless adapters. Newer hardware combines the cable modem, network router, and wireless access point into one device.

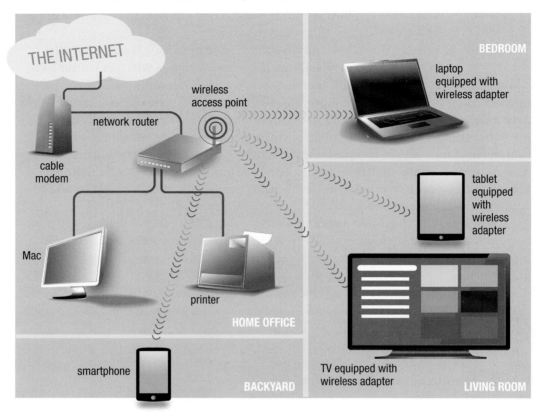

Network Operating Systems

Networks use a few different types of software to function, one of which is a network operating system.

In a network, a **network operating system (NOS)** is installed on the central server. A network operating system includes programs that control the flow of data among clients, restricts access to resources, and manages individual user accounts.

Some popular network operating systems are Microsoft Windows Server in various editions, Open Enterprise Server (OES), Linux Server in various editions, Ubuntu Server, and macOS Server. Another open source network operating system is ReactOS. Note that an NOS typically isn't required for a P2P network. Current Windows operating system versions and macOS have networking capabilities built in, making it easy to set up a basic wireless P2P network in your home or office.

> **Course Content**
> Take the Next Step Activity

6.7 Securing a Network

Whether on a home network or large company network, security is a vital concern today. That's because criminals and malicious hackers can find ways to break into a network to steal data and cause problems. They may locate sensitive financial information, plant viruses that destroy data, or modify settings in ways that cost you time and money. **Hardening** a computer network involves using a combination of hardware, software, and computer-user policies to make the network more resistant to external attacks. Computer security specialists use several steps to harden a network to form layers of protection. For example, a home or business network connecting computers set up to automatically install Windows security updates that contain up-to-date antivirus software running in real time has two layers of protection. In today's home networks, which connect an increasing number of IoT smart devices, network security is more important than ever.

Network security is managed through a combination of techniques involving hardware and software such as a **firewall**, which stops those outside a network from sending information into it or taking information out of it (Figure 6.14). On your home

FIGURE 6.14 **How a Network Firewall Works**
In this intranet, all the devices containing company data are protected behind a firewall, which blocks unauthorized access. Access is controlled by the system administrator.

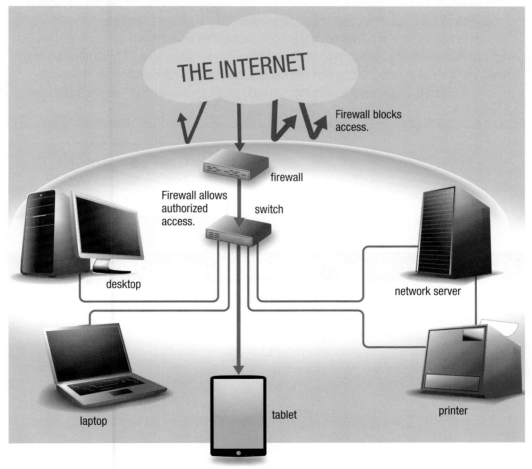

network, you may be the administrator (operating systems such as Windows allow you to set up administrator privileges) and set up your own firewall. A business network requires more complex security measures, because any business needs to make its private network accessible by a variety of people. An employee signs in to the organization's intranet with a user name and password. The intranet is secured by restricting access to authorized users only. For employees who need to access the company's private network from a public location such as an airport or hotel, the company creates a **virtual private network (VPN)** by using a VPN Server. The employee uses a public network (such as the airport's free Wi-Fi) to connect to the VPN Server. The VPN Server establishes the connection to the private network and encrypts all data that passes between the employee's computer and the private network. An extranet, the private network used by businesses, provides secure access to data by vendors, customers, or other authorized users. In business, network security is an in-demand career specialty. Chapter 8 explains computer security in more detail.

Computers in Your Career

Computer security is a hot field for those seeking a career working with computer technology. More and more companies understand that protecting their information and their customers' privacy is essential to their success. Security specialists are employed to assess a system's vulnerability and implement security measures. According to the Bureau of Labor Statistics' *Occupational Outlook Handbook*, Information Security Analysts' responsibilities are continually expanding as the number of cyberattacks increases. This job requires someone with a bachelor's degree for entry-level positions. Additionally, computer analysts and developers will develop new antivirus software, programs, and procedures. Having a security certification may mean you'll make a higher salary than someone without a certification. Consider researching the security certifications most in demand and recognized by employers if you want to pursue this field.

Course Content
Take the Next Step Activity
Ethics and Technology Blog

6.8 Interesting Trends in Networking

Two key technologies of interest in networking circles today are *cloud computing* and *the emergence of 5G*. In fact, the trends are complementary, as both allow computer users more freedom to access what they need to get their work done from anywhere at a lower cost.

> " [5G] has the potential to usher in the fourth industrial revolution — it's that massive. "
>
> —Nicola Palmer, Verizon Chief Network Officer

Cloud Computing

In the past, elements of the network that were kept invisible to users were considered to be in the cloud—that is, software installed on a network server that you simply open and use without having to install it on your computer.

Web 2.0 has ushered in an online cloud where software and services are accessed on the internet (Figure 6.15). In previous chapters, you were introduced to cloud computing as a way to access storage and software applications as services online. This is part of cloud computing, which relies on the concept of Software as a Service, or SaaS. By having applications installed, maintained, and updated outside their walls in the cloud, companies save money.

Some basic cloud services such as Office Online and Google Docs are free for individual users, but more advanced cloud services cater to larger organizations. In 2013, Microsoft released Office 365 to home and business customers. Office 365 is a subscription-based license to the full-featured Microsoft Office suite, indicating the future is subscription software in the cloud for most users. Subscription software such as Adobe Creative Cloud allows you to access your software by signing into a cloud-based account. Such a cloud service has three features that differentiate it from software located on your computer or network:

- It is sold on demand and billed by the minute, by the hour, monthly, or annually.
- People can use as much or as little of a service as they want at any given time.
- The service is fully managed and maintained by the provider. (The end user only uses a computer and his or her internet access to work with the service.)

FIGURE 6.15 **How Software and Data Can Be in the Cloud**
Accessing software and/or storing data in the cloud means your device is the tool that securely connects you to the services and data for which you have subscribed via an internet connection. You are no longer tethered to a specific desktop, laptop, tablet, or smartphone.

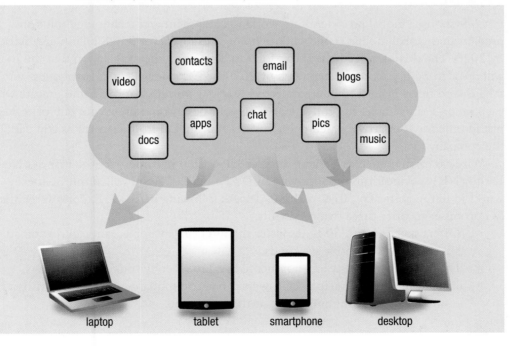

Cloud computing expenditures are a significant focus of IT budgets with a trend toward hybrid cloud deployments. A **hybrid cloud** is created when a company manages some computing resources and data in-house but also has other services provided by outside vendors. For example, a company might use a cloud provider such as Amazon for storing archived data but maintain current data on internal systems.

An emerging trend designed to use cloud computing to address network security challenges for businesses is **Firewall as a Service (FWaaS)**. FWaaS connects an entire organization's network traffic from all locations and all users into a cloud infrastructure, where the cloud's firewall enforces security policies. Businesses are adopting FWaaS to have a consistent security policy that stays up to date, provides access to security experts, and experiences no downtime or deviations in protocols. Using a cloud firewall, a business can free its computer support staff from the work of managing patches and upgrades and constantly assessing security threats. The market for FWaaS is predicted to surpass $2.5 billion by 2024, according to research by Global Market Insights, Inc.

Another interesting trend in networking is called *edge computing*. **Edge computing** uses locally based (close to where the data is generated) data centers to process critical data and then sends batches of the processed data to a cloud provider for storage. Edge computing is typically used as a solution for two types of networks: a network with many connected IoT devices that send a lot of data; and a network in which a reduction in time for data to travel across the network for processing is a critical factor for operations (such as health care, manufacturing, and financial services data). Edge computing allows data to be processed closer to its origination in near real-time rather than sending the data across a long path to a data center in the cloud. Some businesses that have not yet adopted a cloud strategy could also employ edge computing strategies to avoid sending data a long distance to a private data center.

The Evolution of 5G

Cisco Systems, a global provider of networking equipment and technology solutions, predicts that global mobile traffic will reach 366.8 exabytes (or 366.8 billion gigabytes) by 2020. The driving force behind this forecast rests upon the expectation that the IoT will grow to 26 billion connected units by 2020. IoT excludes PCs, smartphones, and tablets, and refers instead to interconnected everyday objects that can communicate with each other. For example, the coffee maker on your desk would brew your cup of coffee just in time for your arrival based upon updates transmitted from your car as you park in the company lot.

Wireless carriers such as Verizon and AT&T wasted no time after the first 5G New Radio (NR) specification was approved in December 2017 and rolled out 5G networks in limited cities in 2018, with 5G devices following in 2019. Experts agree that 5G promises to offer three major benefits:

- Offers speeds up to 50 times faster than 4G (for example, you can download an HD movie in about a second).
- Delivers virtual reality and augmented reality video with no delays or lags.
- Provides greater capacity; cell towers can accommodate more people with more devices all communicating at the same time.

A survey of current news reveals the following 5G predictions:

- A company such as Uber could use a 5G network to help them decide whether to send you a car with a person driving it or an autonomous vehicle. 5G could also help you find a parking spot in a busy city.
- T-Mobile CTO Neville Ray has stated that you could sew a 5G-enabled self-powering wireless sensor into a piece of clothing. "The coat never gets lost and maybe the child never gets lost wearing that coat. They're big ideas, big concepts, and 5G is going to unleash some of those."
- According to a special report on wireless carriers from www.tomsguide.com, "5G could finally make augmented- and virtual-reality headsets more palatable for mainstream users. This is notable because companies such as Apple are reportedly developing augmented-reality glasses to complement—or perhaps even replace—smartphones. 5G would make that possible."

New 5G networks will enable faster connections and a greater number of devices, such as those in a smart home, to connect to networks simultaneously.

Course Content
Take the Next Step Activity

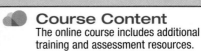

Summing Up

6.1 How Does the World Use Networking?

A **computer network** consists of two or more computing devices connected by a communications medium, such as a wireless signal or a cable. A network provides a way to connect devices, share files and resources, and connect smart appliances to the internet.

Many companies have their own networks, called **intranets**. Intranets are essentially a private internet within the company's corporate "walls." An **extranet** is an extension of an intranet that allows interaction with those outside the company, such as suppliers and customers.

6.2 Exploring Communications Systems

A computer network is one kind of **communications system**. This system includes hardware to send and receive data, transmission and relay systems, common sets of standards (called **network protocols**) so all the equipment can "talk" to each other, and communications software.

Two kinds of signals are used in transmitting data over a computer network: analog and digital. An **analog signal** is formed by continuous waves that fluctuate from high to low. A **digital signal** uses a discrete signal that is either high or low.

If you send data between computers using an analog medium such as a phone line, the signal has to be transformed from digital to analog and back again to digital. Today, most new communications technologies simply use a digital signal, saving the trouble of converting transmissions.

The speed of transmission is determined by two factors: 1) **frequency**, which is the speed at which a signal can change from high to low; and 2) **bandwidth**, which refers to the number of bits per second that can be transmitted. Any communications medium that is capable of carrying a large amount of data at a fast speed is known as **broadband**.

6.3 Transmission Systems

Wired transmissions send a signal through various media, including:
- **Twisted-pair cable**, which is used for your wired telephone connection at home.
- **Coaxial cable**, the same cable used to transmit cable television signals over an insulated wire at a fast speed.
- **Fiber-optic cable**, a very fast system that uses a protected string of glass to transmit data as beams of light.

Wireless transmissions in use today include cellular, microwave, and satellite. All use radio waves to transmit data.

A **cellular network**, like those used by your cell phone or smartphone, transmits signals called **cellular transmissions** by using cell towers. A **microwave** is a high-frequency radio signal that is directed between microwave station towers that are within sight of each other. In 2018, carriers deployed the first 5G networks, bringing potential speeds up to 50 times faster than 4G to the wireless world.

Satellite communication uses space-based equipment for longer range transmissions.

6.4 Communications Standards and Protocols

To allow different devices to talk to each other, the computer industry has developed **standards** that address issues of compatibility. A standard that specifies how two devices can communicate is called a *protocol.*

There are two primary wired standards: 1) **Ethernet**, which specifies that there is no central device controlling the timing of data transmission; 2) **TCP/IP**, in which **packets** are sent and reassembled by the receiver. TCP/IP is the standard upon which internet communications are based.

Wireless networking standards include:

- **Wi-Fi**, which refers to a network that is based on the **802.11 standard**. A location that makes Wi-Fi access available is called a **hotspot**.
- **Long Term Evolution (LTE) standards** prepare mobile phone networks for new technologies, involving areas such as improved bandwidth efficiency and cost controls.
- **Bluetooth**, a network protocol that offers short-range connectivity (3 to 300 feet, depending on a device's power class) via radio waves between Bluetooth-enabled devices such as your smartphone and car. **Bluetooth LE** (low energy) is designed to operate at a lower energy consumption level and a lower cost. Smartphone users may be able to share the internet connection of their device via a cable, Bluetooth, or Wi-Fi with another device such as a tablet or laptop. This process is known as *tethering*, and it is useful when accessing the internet via other means not available for your laptop or tablet. With **Bluetooth 5**, you can play audio on two connected devices at the same time with data transfer speeds up to 2 Mbps over an expanded connectivity range up to 800 feet.
- **Radio Frequency Identification (RFID)**, a wireless technology primarily used to track and identify inventory or other items using radio signals.
- **Near Field Communication (NFC)** is the RFID technology that is used for contactless payments including Apple Pay and Android Pay.
- **Wireless Application Protocol (WAP)**, which specifies how mobile devices such as smartphones display online information, including maps and email.

6.5 Network Classifications

Networks are classified by three characteristics: 1) the size of the geographic area in which the network functions, 2) how data is shared and stored on the network, and 3) how devices in the network are physically arranged and connected to each other.

Types of networks include:

- A **local area network (LAN),** a network where connected devices are located within the same room or building, or in a few nearby buildings.
- A **metropolitan area network (MAN),** a network that connects networks within a city, a university, or other populous area to a larger high-speed network.
- A **wide area network (WAN),** which services even larger geographic areas.

Network architecture relates to how computers in a network share resources. The two major architectures are client/server and peer-to-peer. In a **client/server network**, a **server** computer stores programs and files that any connected device (**clients**) can access. The client/server architecture is considered a **distributed application architecture**. In a **peer-to-peer (P2P) network**, each computer in the network can act as both server and client. A P2P network used on the internet is called an **internet peer-to-peer (P2P) network**.

6.6 Networking Devices and Software

Networking devices include hardware that facilitates the exchange of data from your computer to a transmission medium such as a cable connection, or hardware that enables various devices on a network to communicate with each other. Each device connected to a network is called a **node**.

A modem is the piece of hardware that sends and receives data from a transmission source such as your telephone line or cable television connection. Types of modems include **dial-up**, **cable**, **DSL**, and **wireless**. A **mobile broadband stick** is a USB device that acts as a modem to give your computer access to the internet.

A **network adapter** provides your computer with the ability to connect to a network. A **network interface card (NIC)** is one kind of adapter card. NICs support Ethernet.

A **wireless interface card** functions in the same way as a NIC, except that a wireless interface card uses wireless technology to make the connection.

Wireless access points contain a high-quality antenna that permits wireless devices to transmit data to each other or to exchange data with a wired network.

A **router** allows you to connect multiple networks (or multiple devices if used in a home) in either a wired or wireless connection **(wireless router)**.

A **repeater** is an electronic device that takes a signal and retransmits it at a higher power level to boost the transmission strength.

A **switch** joins several nodes together to coordinate message traffic in one LAN by checking the data in the packets it receives and delivering each packet to the correct destination.

Gateways and **bridges** help separate networks to communicate with each other.

A **network operating system (NOS)** includes programs that control the flow of data among clients, restrict access to resources, and manage individual user accounts.

6.7 Securing a Network

Network security is managed through a combination of hardware and software such as a **firewall**, which stops those outside a network from sending information into it or taking information out of it. **Hardening** a computer network means using a combination of hardware, software, and computer user policies to make the network more resistant to external attack.

A business might use a **virtual private network (VPN)** to protect its network when employees access it remotely.

6.8 Interesting Trends in Networking

The use of cloud computing is expected to continue to grow rapidly in the business world. Cloud computing is where software and IT services are accessed on the internet and are usually billed as a service rather than by application license.

An emerging trend designed to address network security challenges for businesses by using cloud computing is **Firewall as a Service (FWaaS)**. FWaaS connects an entire organization's network traffic from all locations and all users into a cloud infrastructure, where the cloud's firewall enforces the security policies.

Edge computing uses locally based (close to where the data is generated) data centers to process critical data and then sends batches of the processed data to a cloud provider for storage.

By having applications installed, maintained, and updated outside their walls in the cloud, companies save money. The wireless industry started 5G deployments in 2018, with experts at one wireless carrier predicting 5G speed to be up to 50 times faster than 4G.

Terms to Know

6.1 How Does the World Use Networking?

computer network, 148

intranet, 148

extranet, 148

6.2 Exploring Communications Systems

communications system, 148

network protocol, 149

analog signal, 149

digital signal, 149

modem, 150

frequency, 150

bandwidth, 150

broadband, 150

kilobit per second (Kbps), 151

megabit per second (Mbps), 151

gigabit per second (Gbps), 151

terabit per second (Tbps), 151

petabit per second (Pbps), 151

6.3 Transmission Systems

backbone, 152

twisted-pair cable, 152

coaxial cable, 152

fiber-optic cable, 152

cellular network, 153

cellular transmission, 153

microwave, 153

satellite communication, 155

6.4 Communications Standards and Protocols

standards, 156

American National Standards Institute (ANSI), 156

Institute of Electrical and Electronics Engineers (IEEE), 156

Ethernet, 156

TCP/IP, 156

packet, 156

packet switching, 156

Wi-Fi, 157

802.11 standard, 157

hotspot, 157

Long Term Evolution (LTE), 157

Bluetooth, 158

Bluetooth LE (BLE), 158

Bluetooth Smart, 158

Bluetooth 5, 158

piconet, 158

Radio Frequency Identification (RFID), 158

Near Field Communication (NFC), 158

Wireless Application Protocol (WAP), 159

6.5 Network Classifications

6.6 Networking Devices and Software

6.7 Securing a Network

6.8 Interesting Trends in Networking

Projects

Check with your instructor for the preferred method to submit completed work.

Wi-Fi Know-How

Project 6A

To familiarize yourself with how Wi-Fi networks work, watch a video at the web page titled *What Is Wi-Fi?* at https://ODW5.ParadigmEducation.com/WhatIsWi-Fi and another video at the web page titled *How WiFi Works in 4 Minutes* at https://ODW5.ParadigmEducation.com/HowWiFiWorksIn4Minutes. Based on the information you learn from the two videos, prepare a document containing a bulleted list of key facts about Wi-Fi.

Project 6B TEAM

Have your team split into two groups. Each group should visit a place where free public Wi-Fi is available, such as a library, coffee shop, or shopping center. See if you can spot the wireless access points mounted on walls, ceilings, posts, or other structures. How many do you see? (These devices should be apparent by the antennae on the device and are generally mounted high up in the space.) Using your smartphone or other mobile device, connect to the Wi-Fi network. Is it easy to identify and log on to the network? Do you need a password from an employee? Do you have to agree to Terms of Service before continuing to your browser page? If yes, are the Terms of Service easy to read and understand?

Next, conduct a basic internet search to find something like a map or list of movies at local cinemas. Observe the speed as you navigate the internet. Are you satisfied with how quickly the web pages load or is there a lag? If the pages load more slowly than you like, look around the facility to see how many other people might be using the network. Is it possible the network is congested? Consider using a web-based speed measurement tool (such as speedtest.net) to measure the network speed. Meet back with the other half of your team and compare your experiences. Prepare a brief report that summarizes your Wi-Fi experience at each location.

Public Connectivity

Project 6C TEAM

In January 2018, the city of Vancouver in Canada announced an expansion to their free public Wi-Fi network, making it one of the largest free public Wi-Fi networks in North America. Vancouver joins other cities such as Barcelona, Paris, and New York that offer free wireless network connectivity to citizens and visitors. Assume your team has been hired by a local community organization to make a pitch to your city government to fund free outdoor public Wi-Fi for your entire city. Create a presentation with your team's sales pitch to your local city government.

Project 6D TEAM

Your team works at a local property management company that primarily leases commercial space for offices. The CEO of the company has been considering the idea of providing free wireless access in all of the buildings the company manages. To determine the feasibility of this plan, the CEO has asked your team to investigate the benefits and drawbacks of free and open wireless accessibility, including the cost, security, and other factors for its implementation. Prepare a slide presentation for the executive board that supports or discourages this plan, and provide a rationale for your team's position.

Smart Homes of the Future

Project 6E

Some people are already managing parts of their home using their smartphones. You may, for example, have a security system, video doorbell, or smart thermostat that you control from your phone. In the not too distant future, more people will be managing home devices from a smartphone or by using voice commands to turn lights on and off, regulate temperature, monitor a security system, turn on the oven, and check the contents of a smart fridge. Countertops and other surfaces could show live content and updates as you move from room to room. Every appliance and system will be continuously monitored and adjusted to save electricity when not in use, or when you're not home. These technologies are already available. So, what would your ideal smart home of the future do? Go to YouTube and search for videos on smart homes of the future. Watch at least two videos that demonstrate possible technologies that could be used in a smart home. Next, imagine a typical day in your life in a smart home in 2025. Write a one-page movie script of the dialog that might go on between you and your house as the day goes on. For example, your house's first comment of the day might be "Good morning, time to wake up. I'll get the coffee going!" What would be your first request of the day? What other verbal interactions will you have during the day? Submit the URLs of the videos you watched along with your movie script.

Project 6F TEAM

Within your team, consider the smart home of the future and talk about the advantages and disadvantages of living in these connected homes. For example, smart homes will offer many conveniences, but all these devices need an always-on internet connection. How will you be able to run these appliances if the connectivity is down or unreliable? Other issues to consider could include cost, privacy, and the vulnerability of having these interconnected systems subject to a hacker's malware. Prepare a brief presentation with the advantages and disadvantages your team discussed and the conclusions the team reached about each one.

RFID—Cure or Curse?

Project 6G

Radio Frequency Identification (RFID) technology applications have exploded far beyond businesses using tags to track inventory through their supply chains. RFID tags are being used for safety practices, traffic monitoring, healthcare research and monitoring, food quality monitoring, and even tracking waste disposal. But, privacy concerns have arisen about the use of RFID technology. Find and read two articles about RFID from reputable sources. One of the articles should document the benefits of how RFID technology is being used. The second article should address privacy or other security concerns about the use of RFID technology. Create a document that summarizes in a bulleted list what you learned from the two articles; include the URLs for the two articles as well.

Project 6H TEAM

Within your team, compile a list of the benefits of using RFID technology and write a brief statement about privacy and/or security concerns. Each team member should present this information to three relatives or friends, ask each person the following two questions, and then record the answers.
- Do you have any concerns about widespread use of RFID technology?
- How should monitoring and use of data obtained from reading RFID tags be controlled?

As a team, discuss the responses you received to the two questions from all participants. Briefly summarize the participant responses and your team's conclusions about the future of RFID applications in a document or presentation.

Wiki Project—Networking Devices and Terms

Project 6I TEAM

Your team will be assigned to research and write a brief explanation on one or more of the following devices or terms commonly associated with corporate networks: router, switch, bridge, repeater, gateway, firewall, wireless access point, wireless LAN controller, TCP/IP, T1 line, T3 line, T4 line, or VPN. Post your response for the topic(s) assigned to you to the course wiki site. Include an explanation of the device or term, along with examples or images, if possible, to aid comprehension. Be sure to compile a list of references for your sources of information.

Project 6J TEAM

Your team is assigned to edit and verify the content posted on the wiki site from Project 6I. If you add or edit any content, make sure you include a notation within the page that includes the team members' names and the date you edited the content—for example, "Edited by [team members' names] on [date]." Keep a list of references that you used to verify your content changes. When you are finished with your verification, include a notation at the end of the entry—for example, "Verified by [team members' names] on [date]."

Wiki Project—Acceptable Use of Computers and Networks at Work

Project 6K TEAM

Due to ongoing issues regarding the improper use of company-owned computing equipment, the executive board has asked you or your team to draft a new corporate policy on acceptable computer and network practices. You will be assigned to write a section for the new policy on one or more of the following topics:

- Gaming
- Personal email
- Social networking (e.g., Facebook)
- YouTube
- Blogging
- Access to high-quality color printers
- Company-provided computing equipment (e.g., laptop, tablet, smartphone)
- Internet surfing or shopping
- Texting and video chatting

When you have finished writing your assigned section, post your topic and its accompanying text to the course wiki site.

Project 6L TEAM

Your team has been asked to comment on the content posted on the wiki site from Project 6K. Post your commentary followed by a notation that includes the team members' names and the date you commented—for example, "Comment by team members' names] on [date]."

Class Conversations

Topic 6A   Can you see me now?

Location-Based Services (LBS) available for smartphones have opened new doors to managers who want to keep a watchful eye on their mobile workers. Service agreements are available with a company's cellular provider to log employees' movements, including tracking when the employee has entered or exited a predefined region (called *GeoFencing*) and sending an alert when a specified speed limit has been exceeded. While LBS has obvious safety benefits (such as locating a mobile worker who may be lost or in need of assistance), is there a point at which an individual's right to privacy is at risk with this technology? If yes, where would you suggest this tracking technology has the potential to invade an employee's privacy? Assume you are a manager at a company that subscribes to LBS. How will you address any privacy issues raised by your employees about the data you will receive from the mobile device?

Topic 6B   Should everyone open their wireless networks to everyone else?

"The Open Wireless Movement is a coalition of internet freedom advocates, companies, organizations, and technologists working to develop new wireless technologies and to inspire a movement of internet openness. We are aiming to build technologies that would make it easy for internet subscribers to portion off their wireless networks for guests and the public while maintaining security, protecting privacy, and preserving quality of access."

—openwireless.org

The Open Wireless Movement wants a future in which the internet is entirely open to everyone in urban environments. To achieve this goal, all internet subscribers would have to allow open access to their own wireless networks. Visit openwireless.org and explore the reasons, myths, and facts about open wireless. Would you consider opening up your home Wi-Fi network to be accessed by strangers? Why or why not?

Topic 6C   Is network security threatened with a BYOD policy?

BYOD (which stands for bring your own device) policies allow employees to use personally owned mobile devices for work. This means that employees are gaining access to company networks and possibly storing sensitive corporate information on their personal devices. Network security specialists see a smartphone, tablet, or laptop as an easy portal through which hackers can gain access to a corporate network when users have lax security on their personal device. Should a company allow employees to use their own personal devices for work? Why or why not? If a company has a BYOD policy, how might the organization harden their corporate private network?

The Social Web
Opportunities for Learning, Working, and Communicating

What You'll Accomplish

When you finish this chapter, you'll be able to:

7.1 Explain the social web phenomenon and its impact on how our society functions.

7.2 Examine the past, present, and future of social technology and how our lives have changed through this development.

7.3 Identify a blog and explain the uses of blogs in today's workplace and personal settings.

7.4 Describe the development, growth, and trends of social networking.

7.5 Explain how social bookmarking works and identify three different services.

7.6 Identify the role of wikis and explain how people are using wikis in the social web.

7.7 Explain the role of media sharing and provide examples of how it is being used.

Why Does It Matter ?

Social networking sites, such as Facebook, have seen phenomenal growth in the past several years. At the end of 2017, Facebook had more than 2.13 billion active users. Facebook has experienced a tremendous increase in mobile users of its site, growing to its current user base from only 750 million in 2011. If Facebook were a country, it would be the world's largest, with more than six times the population of the US. But connecting with friends on social networking sites is only one aspect of a trend toward a more collaborative online environment. Social technologies are also being used by businesses to help employees connect with each other and communicate with customers, by nonprofit groups to document trends and promote social causes, by schools to involve students in collaborative projects and interactive learning, and in many other settings. You are part of a revolution in the way that people share and collaborate, and by choosing to embrace that revolution, you may reap benefits in many areas of your life.

 The online course includes additional training and assessment resources.

From social networking sites that contain networks of friends to collaborative wikis, blogs, and media sharing sites, social websites represent a revolution in how people connect, learn, and work together.

Course Content

Take a Survey

Technology in Your Future Video

7.1 The Social Web Phenomenon

You may belong to or have visited pages on websites such as Facebook, Twitter, YouTube, Pinterest, Google+, or Tumblr, or have read about these sites in the news. All of these are social sites, where people go to share their thoughts in text, video, and/or photos. Together with a wide variety of other social websites, these sites form the **social web**, a revolution in how people connect with each other, how news is delivered, and how our collective knowledge is formed. Social sites and the tools they offer create a vehicle for a two-way dialogue between people and groups, rather than a one-way communication from the media to the public, stores to customers, or teachers to students. Of course, there is a downside to social sites, as well. Keeping up your online social image can cause stress, loss of sleep, and envy. Cyberbullying on social media can have an impact on an individual's self-esteem and cause depression or worse. Time spent on social sites can take away from other activities and interactions in real life.

> "Facebook was not originally created to be a company. It was built to accomplish a social mission—to make the world more open and connected."
>
> —Mark Zuckerberg, founder of Facebook

The social web is still evolving and defining itself, and as such is likely to include more types of websites and services than you think. Any site that allows users to interact with each other and share information or content can be considered, at some level, to be social. A website that allows you to share contacts and build a network of friends is a **social networking site**. Services that allow you to share media are **media sharing sites**. Online dating services and special interest sites, such as those about sports or genealogy, when they allow interaction and communication among members, are social sites. Social sites such as Pinterest also allow for sharing of visual images, articles, and collections of ideas related to travel, cooking, craft projects, and any other hobby or topic you may be interested in. The content of most of these sites is driven almost entirely by the members, though the site owners devise and maintain the infrastructure, communication tools, and rules for behavior.

7.2 Social Technology Comes of Age

In Chapter 2 you read about the interactive web associated with user involvement and collaboration. This collaboration happens through interactive web services such as Wikipedia (an online encyclopedia) and Facebook. These web services provide users with a way to share information, exchange ideas, and add or edit content in collaborative articles. The social web is one of the most publicized and successful uses of the internet.

How the Social Web Was Born

Since the early days of the internet, when it was used only by government and educational researchers, through the first few years after the internet became available to the general public, people have been interacting online through tools such as discussion boards and email, but the social web took that interaction much further. The concept of online interaction came into its own in 2001, and understanding the trends it describes is important in understanding how the social web came to be. In the late 1990s, the open source movement allowed individuals to contribute to the source code of free software, such as the Linux operating system. **Open content** is to the social web what open source was to software development—it means that people can freely share their knowledge about topics in online collections, such as Wikipedia.

In 1997, at about the same time that the open source and open content movements were growing, one of the first true social networking sites, SixDegrees.com, appeared. Although other sites already existed that allowed users to create **profiles** with information about themselves and compile lists of friends, SixDegrees was one of the first sites that combined the use of profiles and searchable **friends lists** into one service. Social networking was born.

SixDegrees failed, perhaps in part because its concept was ahead of its time. In 1999, online journals called **blogs** (a term created from *web + log*) began to surface online, facilitated by blogging sites such as Open Diary and Blogger. Blogs

Social sites allow users to share ideas and content online.

became one of the key tools for online social interaction. A few more social networking sites appeared throughout the late 1990s, many incorporating a blogging component, until, in 2003, the phenomenon exploded with sites such as MySpace, Flickr, Facebook, and LinkedIn all launching within months of each other. Google+ was launched in 2011, amassing over 10 million users within the first two weeks. Figure 7.1 presents a timeline of social networking sites.

FIGURE 7.1 **Launch Dates of Social Networking Sites**

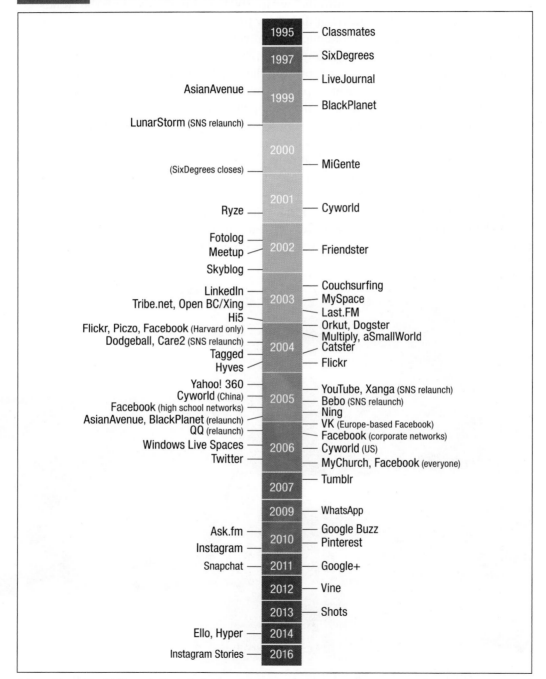

Adapted from "Social Network Sites: Definition, History, and Scholarship" by Boyd and Ellison, *Journal of Computer Mediated Communication*.

The social web makes it possible for streams of information and connections to travel across the globe.

Overview of Social Technology Today

Today, the social web can be accessed by a variety of devices, such as smartphones and gaming devices. These devices allow over 2.5 billion people to connect with their social sites to post text, video, and photos and interact with their friends on the go. Certain social media services, such as Snapchat, are accessible only via apps on mobile devices—their content cannot be viewed using a browser.

If you perform a search using the terms "social web," you will see the incredible variety of social sites and applications available to you. The functions of these sites and applications are as varied as the people and organizations who use them. Social causes use the social web to muster support in times of crisis, such as raising money to help victims of natural disasters. Businesses use social marketing, driving their branding and sales messages to the public by participating in all kinds of social networking sites. Politicians hold online dialogs on political and social issues to gather votes and support. Artists create musical and visual pieces by sharing media and building new pieces of art collaboratively. Political demonstrators all over the globe have used social media to organize their efforts and communicate about incidents and developments.

The social web is growing and evolving rapidly, with changes happening daily. Functions and features of the different social websites overlap, making it challenging to define the technologies precisely. However, grouping them into the following broad categories provides a way to examine them and understand their value in our digital world:
- blogging
- social networking
- social bookmarking
- wikis
- media sharing

Each of these categories of social media is explored in this chapter in terms of how the technology works, who uses it, and for what purposes.

Khan Academy is just one example of a site that provides students with videos, practice exercises, and assessments for free.

The Future of Social Technology

By the time you read this chapter, social technology will have changed. That's one of the most exciting things about web content driven by the masses: it morphs very quickly because anybody can suggest an idea that becomes the next great trend, rather than having trends dictated by businesses or the media. Still, it's possible to speculate about some future directions for the social web that are already emerging.

One predicted trend is the ability to carry your **identity** (the profile you create when you join a service) with you from site to site. There will be a connection among all the social sites you now use separately. You will have one set of friends who have access to your page and one set of **preferences** (such as privacy settings). Currently, you can see the potential of this kind of system in the ability to sign up for or into multiple sites using your Facebook credentials. Social media and messaging are available on mobile devices. Social messaging has evolved as a component built around social networking platforms such as Facebook Messenger. Businesses can use social messaging to connect with customers.

> "Social networking in the enterprise will break down. . .barriers and provide equal access to information across levels and job functions."
>
> —Luosheng Peng,
> CEO, Gageln

Another trend is the ability to gather together content from a wide variety of services. For example, a service like HootSuite allows you to create categories of information and post updates on a variety of services, such as Twitter, Facebook, and more, in one action.

It's been suggested that in the future, all websites will have social networking features, and in fact, that movement is already underway. For example, Outlook.com, which started as an email and calendar service, now allows you to build a friends network, as do some bookmarking sites. You may also be able to use your social media account to log in to interactive features of other media websites, such as the comments feature for online newspapers. Social media has become a real-time reporting tool. In 2017, news about Hurricane Maria in the US spread quickly as Puerto Ricans used social media to raise funds to aid those in distress. Recent changes in government regimes in the Middle East spread through the use of social media sites such as Google Crisis Response, which allows people to get news updates and share emergency-service data in critical times. What could that mean to the way we share knowledge, do business, become aware of global social causes, and report the news?

HootSuite is a social relationship platform that allows users to control their various social media profiles from one secure dashboard.

Course Content
Take the Next Step Activity
Ethics and Technology Blog

7.3 Blogging: The Internet Gets Personal

A blog is an online journal that anybody can use to express ideas and opinions online. A blog may be focused around a particular topic, such as animal rights, or simply be a random collection of personal thoughts. Blogs can contain text, images, videos, and links to other online content. Most sites that host blogs allow bloggers to set a level of privacy that determines who can view their content.

People who read blogs can post comments about blog entries. Blogs and responses to them are typically listed in reverse chronological order, with the most recent post at the top.

How Blogging Has Evolved

Blogging began on a small scale in the mid-1990s but gained popularity in 1999, when blogging tools became more generally available to the public through blog hosting services such as LiveJournal and Blogger.

Over time, blogging has moved into many online settings. Today you may create your blog on a blog hosting site, or you may post to a blog that is part of a social networking site such as Facebook, Tumblr, or Bebo. Companies often host blogs on their websites where they can share information with their customers and listen to their customers' opinions through product reviews. Experts from sites such as ZDNet and The Huffington Post write blogs to help those sites extend their readership.

Blogs are incredibly popular, but it's interesting to consider some of the consequences of people posting their opinions and thoughts for millions of others to read. Blogs have been the subject of lawsuits when posts slander another person or organiza-

Commenting on technology trends is a popular use of blogs.

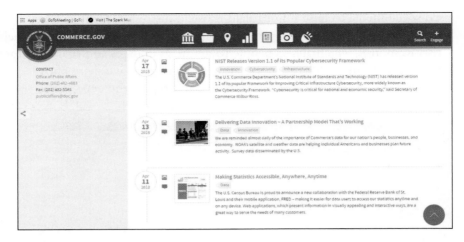

tion. Some governments have cracked down on political blogs that challenge government policy. Job seekers have begun to realize that employers often read (and consider in their hiring decisions) personal information applicants have posted online.

Computers in Your Career

If you have an interest in journalism as a career, you'll find that your chosen field is undergoing major changes. Journalism once provided a "one speaking to many" model, but blogging has brought a new dynamic of two-way communication. Old models of print newspapers and magazines are being challenged, but the internet is also bringing new opportunities. For example, mobile phone journalism (MoJo) offers the ability to report instantly from the field via mobile phones. According to John S. Carroll, former editor of the *Los Angeles Times*, "Journalism…is now a conversation with millions of participants, which gives us access to new facts and new ideas. Thanks to hyperlinks, you can write accordion-like stories that can be expanded to match each reader's degree of interest. The journalism of the future will be flexible, making fluid use of video, audio, and text to tell stories as they can best be told." Journalism is certainly a career that is undergoing dramatic change; still, future opportunities to take advantage of new media and content choices continue to make this an intriguing career choice.

The Many Uses of Blogs

Many blogs are simply personal journals chronicling a person's day or opinions. But today, blogs have also taken on a role in reporting news stories and in providing a soapbox for experts—or *pundits*—on topics from politics to the environment. Governments of various countries host official blogs where citizens can voice their opinions. People who spot technology and social trends find huge audiences for their blogs, as do entertainers such as Bill Maher and other TV and movie stars. An interesting blog statistic is that every half second somewhere in the world, a new blog is created.

A fascinating aspect of blogging is the use of the social web during crises. In 2017, for example, the international aid organization Oxfam produced a blog about the extreme poverty in Puerto Rico

> If you make customers unhappy in the physical world, they might each tell 6 friends. If you make customers unhappy on the internet, they can each tell 6,000 friends.
>
> —Jeff Bezos, founder of Amazon.com

exacerbated by Hurricane Maria, helping to generate help for its citizens. Social networking has also exposed the reality of conditions in impoverished and politically unstable countries despite government attempts to repress information.

Sites such as Twitter (referred to as **microblogging** or **social journaling** sites) focus on small details of everyday life by allowing users to create brief posts (280 characters maximum on Twitter) about their daily thoughts and activities. While microblogging sites may seem, at first glance, to focus on unnecessary minutiae, they can be used to create change both locally and globally. For example, Twitter established hope140.org to promote social movements around the nation and the world. The popularity of microblogging is shown in its statistical growth. As of January 2018, there were 330 million active users, and over 500 million tweets are sent every day. Eighty percent of Twitter users access the service using a mobile device.

With so much content being constantly created on social media, users needed a system they could use to organize and search for it. Because the pound symbol (#) was already in use in several programming languages, in 2007 it was suggested that this symbol be used as a **metadata** tag to organize topics on Twitter. Metadata is essentially data about data. In 2009, Twitter officially adopted the system by linking search results to terms preceded by the # symbol, officially terming them **hashtags**.

Today, hashtags are used across a wide variety of social media platforms and serve many purposes. For example, they are used by companies to track what people are saying about them, by groups of friends or colleagues to compile and share photos or data from a particular event or project, and by social activists to generate buzz about a cause.

Bloggers organize their content in a similar way: they assign **tags** to particular posts or chunks of content. This allows readers to search a blog and quickly find out what a blogger said about a particular topic. Although many blogs still focus on the written word, others embrace alternate communications media as well. The following are examples of types of blogs that keep us connected:

- **Artblogs** share artwork including photos, drawings, paintings, and/or music. Some artblogs even share creative graffiti.
- **Sketchblogs** focus on illustrations and can be as simple as photos of handwritten personal journal entries that include informal doodles and sketches, or as ambitious as formal galleries presenting and marketing professional work.
- **Photoblogs** include entries that run the gamut from informal personal snapshots to galleries of professional work intended for sale. Pinterest is a popular site you can use to share photos from a variety of sources that relate to a theme or project.
- **Video blogs**, sometimes called *vlogs*, allow people to share their content in real time (a technology called *streaming*). Sites such as JustInTV and LiveStream were early providers of streaming video blogs. This technology has begun to influence how news becomes news.
- **MP3 blogs** such as Buzzgrinder began as a way for musicians to post their music for others to download. Along with Facebook and YouTube, such venues are becoming major outlets for bands wishing to make their music available to a larger audience.

Playing It Safe

Many sites that host blogs provide behavioral policies, as well as blog moderators, to keep contributors from acting abusively towards others. Other sites are run with little supervision and therefore are open to more risk for abuse by other bloggers, often in the form of cyberbullying. If you decide to create a blog, you should choose a site that offers the support and control that fits your style and comfort level.

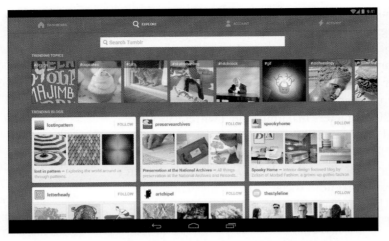

Tumblr is a combination microblogging and social networking site whose customizable format allows users to share text, photos, quotes, music, and videos.

☁ **Course Content**
Take the Next Step Activity
Ethics and Technology Blog

7.4 Social Networking

A social networking site typically includes a blogging-like feature, such as Facebook's status updates, and the ability to share media, but what has traditionally differentiated social networking sites from other social sites is the ability to share contacts and build a network of friends.

However, with sites such as YouTube (a media sharing site) and Outlook.com (traditionally an email service that has recently grown to include some social features) now incorporating a way to build a list of friends, it's hard to identify what a social networking site is anymore. The lines get blurred as the ability to socialize is being built into many different types of websites.

The Amazing Growth of Social Networking

Online socializing began in the late 1970s when Usenet, ListServ, and the bulletin board system (BBS) appeared. These morphed from gathering places for those working on special government research projects to gathering places for those with shared interests.

In the 1980s, early internet service providers such as CompuServe incorporated next-generation bulletin boards called *discussion forums*. Discussion forums encouraged interaction with posts and responses displayed in discussion series called *strings*. Classmates.com came along in 1995 with a focus on providing virtual, online class reunions. America Online (AOL) pushed social boundaries even further with its use of searchable member profiles, that allowed users to connect with each other. This engendered a sense of online community, which was fundamental for the social networking phenomenon to come.

In 2002, Friendster became one of the first social networking sites to catch on with the general public. (The site's popularity in the US dwindled fairly quickly, but it is still popular in Asia.) Friendster was followed in 2003 by LinkedIn, which has a focus on career networking, and MySpace. In 2004, three Harvard University students launched Facebook as a college-student-only service. The site opened to the general public in 2006. Google+ launched in 2011, introducing a new social service available to Android mobile devices.

Today the social networking landscape is rich with a variety of options that extend well beyond Facebook. Some of the other popular social networks include LinkedIn, Pinterest, Instagram, Reddit, Nextdoor, and MySpace. Many other sites not traditionally considered to be social also include social features, such as profiles, friends lists, and media sharing capabilities.

Studies show that social networking has had a real impact on how people in our society relate to those around them. A Pew Research project examined how technology impacted users' social behavior, including relationships, social support, community, and involvement with politics. The research found that social networking is used to maintain existing, strong social connections as well as to revive relationships from years earlier. The research also showed that Facebook users are more trusting and have more close relationships than those who do not use the site.

> " Asking about using social media to advance your career is the wrong question. Think instead about advancing the cause, the company, or the profession. "
>
> —Rosabeth Moss Kanter, Harvard Business School

Interesting Trends for Social Networks

More than 2 billion people use social networks as a part of daily life. As social networking tools continue to be incorporated into traditionally non-social sites and access to social networks becomes increasingly mobile, it is expected that social networking will become an even bigger part of our lives. The future is anyone's guess, but we can cite some of the current trends in social networking.

One established and still growing trend in social media is the use of social networking sites as marketing tools. Businesses use the information you provide on your social media profiles to find out more about who you are as a consumer and whether you might be interested in purchasing their products or services. Over one million websites have now become integrated with Facebook and other social media sites, such as Pinterest, by including a Like button or a Pin It button on their product pages. This allows users to share content directly from

Playing It Safe

People typically want a lot of friends, but having too many friends on a social networking site can be problematic. Be cautious of allowing friends of friends to have access to your page. Remember that a friend of a friend is often a stranger. Protect your privacy accordingly.

Facebook began as a social networking site for college students. Today, even small businesses have Facebook pages.

a business's website on social media, essentially providing free publicity for the company and its products.

Snapchat's My Story feature allows you to post a set of photos. Several new services offer the same opportunity to post photos that create a story, such as Instagram Stories. Rather than posting a single photo, you can post a collection of photos to be played as a slideshow. In 2018, Instagram Stories had over 250 million users.

One of the latest trends in the social web is **disappearing media**, a movement driven by mobile apps such as Snapchat,

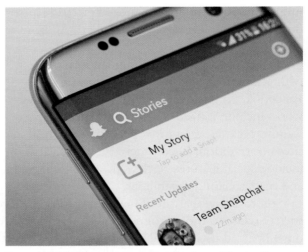

Snapchat My Story is a select collection of "snaps" (photos and/or videos) captured by a user in the last 24 hours that play in the order taken. After 24 hours have passed, the "snaps" automatically disappear.

which allows users to send photos and videos with text captions to a select group of people. Once opened, the images disappear (and are deleted from Snapchat's servers) within a few seconds. As of 2018, Snapchat's users were uploading more than one million photos and videos every day.

Athough social media have become a valuable part of many people's daily lives, certain aspects of these sites—such as the tendency of certain users to "overshare"—can become annoying and tedious. To avoid user burnout, many social media sites are now offering ways for members to turn off content from certain other members without those people knowing about it. Features such as Twitter's Mute option and Facebook's Hide option allow users to privately customize their experience by controlling which of their friends' posts they see or don't see. Because these options improve user experience while still allowing members to be tactful and sensitive to their friends' feelings, this trend will likely continue to grow. More social networking sites are expected to add similar features in the near future.

Another trend that has surfaced over the last few years is increased usage of social networks by older people. Even though social networking is popular with young people, the real growth in social networking has occurred among users 50 and older, and even 65 and older. The social networking phenomenon continues to change the way generations use social media technology. Figure 7.2 illustrates how the use of social networking is continuing to grow among adults.

Mobile use of social networks (so-called **social mobile media**) continues to grow. People are using mobile phones to access their social networking pages, upload pictures and videos they

More people are using their cell phones to communicate with others through social networks.

FIGURE 7.2 **Growth in Social Media Use**

The percentage of US adult internet users who use various social media sites from their cell phones continues to grow.

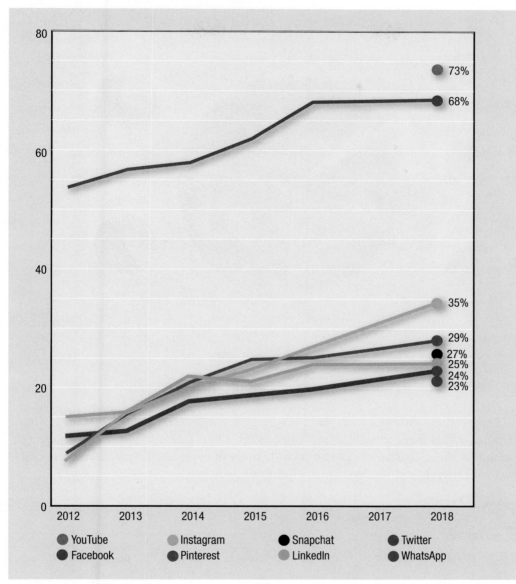

YouTube Instagram Snapchat Twitter
Facebook Pinterest LinkedIn WhatsApp

Source: Pew Research Center *Social Media Use in 2018* as of January 10, 2018. Note: Pre-2018 telephone poll data is not available for YouTube, Snapchat, or WhatsApp.

capture with their phones, tweet thoughts, and comment on friends' posts. Over 73% of smartphone owners access YouTube on their phones. In addition, 68% use their smartphones daily to access Facebook to update their statuses, post videos or pictures, or comment on a post from a connection or friend. As shown in Figure 7.3, time spent on social media sites, and by extension mobile devices, adds up to years.

Social networks are reaching beyond online communication to connect with events in our offline lives. Sites such as Socializr, Punchbowl, and Evite have become popular for organizing social events. People post events, send RSVPs, and later post event photos, which they can pull onto the event site from sites such as Facebook or Flickr.

FIGURE 7.3 **Time Spent on Social Networks in a Lifetime**
The time the average person spends on social sites translates to years over a lifetime.

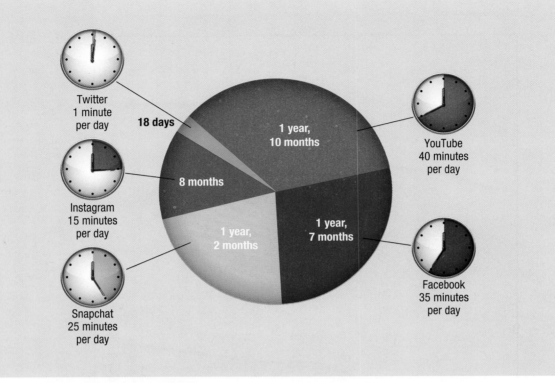

Twitter
1 minute
per day

18 days

Instagram
15 minutes
per day

8 months

Snapchat
25 minutes
per day

1 year,
10 months

1 year,
2 months

1 year,
7 months

YouTube
40 minutes
per day

Facebook
35 minutes
per day

A trend that continues to expand, involving services such as Groupon and Living Social, provides users with discount coupons/vouchers that may be delivered through a mobile device. These services also provide users the opportunity to share these coupons with friends and receive benefits for doing so.

Computers in Your Career

Many people use social tools to market themselves to employers. They use networking sites such as LinkedIn to connect with other professionals. Others use social tools to post comments on blogs related to their career interests to create online credentials that potential employers may notice. Some produce podcasts or upload videos about their areas of expertise to media sharing or job-posting sites to get noticed, or they post online portfolios on their own websites. Next time you're looking for work, consider using the social web to help you land a job.

Course Content
Take the Next Step Activity

Though the social web is in flux, one thing everybody can agree on is that the web contains a huge number of websites and a wealth of content. One form of social networking that helps users organize and recommend content to each other is social bookmarking. Using sites such as Symbaloo, Digg, Folk'd, Reddit, and StumbleUpon, people can share online content with individuals and groups and organize that content in personal libraries.

> " Social bookmarking has become a phenomenon in the last couple of years. As more individuals join these social networks, the news and what is deemed important is now driven by consumers—a fundamental shift on how information was prioritized in the past. "
>
> —Michael Fleischner, Internet marketing expert

How Social Bookmarking Works

Social bookmarking allows you to make note of online content in the form of tags called **bookmarks** and share those bookmarks with others. The technology uses metadata. Metadata describes the location or nature of other data, allowing software such as a browser to organize and retrieve that data easily.

Bookmarking sites save bookmarks as tags rather than saving links in folders, as the Favorites features of some browsers do. As you learned earlier, a tag is a keyword assigned to information on the web that is used to locate and organize content references. Because you can sort through and organize tags, social bookmarks are much easier to search and catalogue.

Above: Flipora provides you with recommendations of sites that you'll love based on your interests.
Right: The web is riddled with logo links like these that allow you to connect with bookmarking and other types of social sites.

Social bookmarking sites such as Pearltrees allow you to compile, organize, and share online content.

A Wealth of Social Bookmarking Services

Today you will find tools on many sites that allow you to bookmark them instantly. Look for logos for services such as Reddit, StumbleUpon, and Pearltrees on your favorite website, or locate a Share icon that, when clicked, displays a variety of tools that allow you to share your recommendations.

> ● **Course Content**
> Take the Next Step Activity

7.6 Wikis

The social web isn't only for swapping personal stories or photos. **Wikis** provide a way to share knowledge about every topic under the sun in the form of online visual libraries, encyclopedias, and dictionaries. Wikis enable people to post and edit content in a way that creates a living network of knowledge to which anybody can contribute.

What's a Wiki?

According to http://wiki.org, a wiki "is a piece of server software that allows users to freely create and edit web page content using any web browser." The wiki technology supports hyperlinks and enables users to create links between internal pages. Wikis allow users to not only edit the content but change the organization of that content as well. Wiki content can then be searched by users to find the information they need.

> 66 The internet has transformed the educational landscape, giving students more scope to access information and offering them the opportunity to collaborate in research projects online. 99
>
> —Aleks Krotoski, journalist

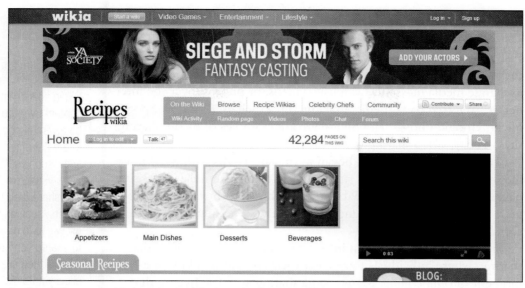

A wiki can be a valuable learning tool that can be integrated into courses for students of all ages.

The wiki "open editing" model encourages anybody and everybody to contribute his or her experience and knowledge in text, audio, or video format. When you use the edit feature in a wiki, it opens the content as a document that you can modify. You can add and edit text or graphics and insert links to other documents. You then save your changes so that others can view and edit the updated document.

The open editing model can make the accuracy of wiki content harder to verify. Some sites, such as Wikipedia, have systems in place to monitor posts and edits and note where additional clarification or authentication is needed.

Who's Using Wikis?

You can use wikis to coordinate projects, trips, and parties, or build online repositories of shared information in the form of encyclopedias or dictionaries. Authors, artists, collectors, journalists, educators, scientists, researchers, technologists, and business people are making use of wikis to collaborate on creative works and build business policies and procedures. People who share interests are using wikis to build content communities. In business, where companies often have to get client approval of designs or campaigns, wikis help streamline the process and keep everybody in the loop.

Here are some interesting uses of wikis you might want to check out:
- Memory-Alpha.org is a wiki where anybody can contribute to and edit an encyclopedia about all things Star Trek.
- Wikitravel.org is an open content travel guide with advice and information from thousands of travellers.
- Wiktionary.org is an open, web-based dictionary that provides definitions, pronunciations, and the history (etymology) of words contributed by users.
- Teampedia.net is a collaborative encyclopedia of team-building activities, resources, and tools.

What interesting wikis can you find online?

> **Course Content**
> Take the Next Step Activity

7.7 Media Sharing

Just as people want to share stories in blogs and knowledge in wikis, they also have the need to share files. When you view, download, or exchange media files, such as music, videos, or photos, over the internet, it's referred to as **media sharing**. Sites such as YouTube, Spotify, Flickr, Google Photos, and Slideshare are examples of media sharing sites. Today, video sharing is still the most popular sharing trend.

The media sharing trend began in the late 1990s with MP3 music file sharing through services such as Napster and Gnutella. MP3 sites have gone through some legal challenges because they have distributed the work of musicians and artists freely, sometimes violating copyright protections.

Sites such as YouTube tap into the grassroots version of media sharing, where individuals freely post their content in a bid for a moment of online fame. Many media sharing sites use **live media streaming**, a technology that allows them to send the content over the internet in **live broadcasts**.

How Media Sharing Works

You can easily create your own media and share it with others online. Some sites, such as WeVideo.com, allow you to record, edit, and share your media right from the site; others require that you record it offline using a camera or voice or video recorder. Media that you create in the form of a digital file can be shared as an email attachment, sent through instant messages, or posted on sites for people to download.

> "At CUNY's Graduate School of Journalism . . . we just told the students that they no longer need to commit to a media track—print, broadcast, or interactive. We believe this is the next step in convergence. All media become one."
>
> —Jeff Jarvis, *Columbia Journalism Review*

Media is shared on many social networking sites and blogs, such as Instagram and Tumblr, in web-based communities, on social bookmarking sites, and on more specialized media sharing sites, such as Flickr and YouTube.

Many media sharing sites like YouTube are available in web browsers and mobile apps.

SoundCloud, launched in 2008, enables users to stream, record, share, provide feedback on, and remix music and other audio.

VoiceThread is an interesting example of a site that combines media sharing with the ability to hold a conversation about the media. You can even record voice comments from your computer or any phone, including cell phones, and navigate the site through voice commands.

How People Are Using Media Sharing

People are using media sharing in a variety of ways. Some artists are promoting their work by sharing it in online portfolios. People collaborating on projects, such as designing a website, may share media in environments that allow each person to comment on or annotate the media file. Some media sharing sites and software allow you to build personal playlists of the media you find online.

A popular activity is to post product reviews or tutorials; for example a video, showing features of new cell phone models. Businesses and nonprofit organizations can use services such as Radian6 from Salesforce.com to search the internet and find content related to a theme. For example, they can find posted videos that relate to their products or brand so they can learn what their customers think of them and make changes and updates accordingly.

Sites such as FriendFeed allow you to import playlists from several services and even let you share your content through Twitter and Facebook. You can also post your customized list of media content on your own website or blog.

> ◗ **Course Content**
> Take the Next Step Activity
> Ethics and Technology Blog

Review and Assessment

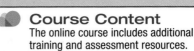

Summing Up

7.1 The Social Web Phenomenon

Social networking sites are websites where people go to share their thoughts in text, video, or photos. Any site that allows users to interact with each other and share information or content can be considered, at some level, to be social. Services that allow you to share media are called **media sharing sites**. Though it provides many opportunities for connection, participation in social sites can also create stress, anxiety, and other negative effects.

A wide variety of social websites such as Facebook and Pinterest form the **social web**, a revolution in how people connect with each other, how news is delivered, and how our collective knowledge is formed. The content of most of these sites is driven almost entirely by the members, though the site owners devise and maintain the infrastructure, communication tools, and rules for behavior.

7.2 Social Technology Comes of Age

The social web is one of the most publicized and successful examples of Web 2.0. **Open content** is to the social web what open source was to software development—it means that anybody can freely share their knowledge about topics in online collections, such as Wikipedia. Although other sites already existed that allowed users to create **profiles** with information about themselves and compile lists of friends, SixDegrees was one of the first sites that combined the use of profiles and searchable **friends lists** into one service, resulting in the birth of social networking. Online journals called **blogs** (a term created from *web + log*) began to surface online. Blogs became one of the key tools for online social interaction. In 2003, the social phenomenon exploded with sites such as MySpace, Flickr, Facebook, and LinkedIn all launching within months of each other. Today, the social web has expanded to be accessible by a variety of devices, such as cell phones and gaming devices. These devices allow people to connect with their social sites to post text, video, and photos and interact with their friends on the go.

The social web is growing and evolving rapidly, with changes happening daily. One predicted trend is the ability to carry your **identity** (the profile you create when you join a service) with you from site to site. Another trend is the ability to gather together content from a wide variety of services. Social media has also become a real-time reporting tool. It's been suggested that in the future all websites will have social networking features.

7.3 Blogging: The Internet Gets Personal

A blog is an online journal that may be focused around a particular topic or may simply be a random collection of personal thoughts. Blogs can contain text, images, videos, and links to other online content. Many blogs focus on the written word, but there are also blogs that use other kinds of content, such as **artblogs, sketchblogs, photoblogs, video blogs** (also know as *vlogs*), and MP3 blogs. People who read blogs can post comments. Blogs and comments are listed in reverse chronological order, with the most recent post at the top.

You may create your blog on a blog hosting site, or you may post to a blog that is part of a social networking site. Companies often host blogs on their websites, and blogs have taken on an important role in reporting news stories. Sites such as Twitter that use brief comments are called **microblogging** sites. Because of the quantity of content on social networking sites, **metadata** tags called **hashtags** (topic names preceded by the # symbol) are being used today to organize content and make it easier to search. Blogs also employ the use of **tags**, which are labels that make it easier for readers to find what they are looking for within a particular blog site.

7.4 Social Networking

Social networking sites typically include a blogging-like feature, such as Facebook's status updates and the ability to share media, but what differentiates them from other social sites is the ability to share contacts and build a network of friends. Many websites that were not traditionally considered to be social now include social networking features such as profiles, friends lists, and media sharing capabilities.

Social networking trends include the increased use of social media as a business marketing tool, the emergence of **disappearing media**, higher rates of social media use by older people, and the rapid adoption of **social mobile media**. The social web is also creating a plethora of other social and business opportunities including sites such as Instagram Stories that allow for collections of images.

7.5 Social Bookmarking

Social bookmarking helps users organize and recommend content to each other. This technology uses metadata. Bookmarking sites save users' **bookmarks** as tags. Because you can sort through and organize bookmark tags, it's easy to use them to search for and catalogue information.

7.6 Wikis

Wikis are a way to share knowledge about any topic in the form of online visual libraries, encyclopedias, and dictionaries. Wikis enable people to post and edit content in a way that creates a living network of knowledge to which anybody can contribute.

7.7 Media Sharing

Viewing, downloading, or exchanging media files, such as music, videos, or photos, is referred to as **media sharing**. Many media sharing sites use **live media streaming**, a technology that allows them to send the content over the internet in **live broadcasts**.

Media is shared on many social networking sites and blogs, in web-based communities, on social bookmarking sites, and on more specialized media sharing sites such as Instagram and Tumblr.

Terms to Know

7.1 The Social Web Phenomenon

social web, 182
social networking site, 182
media sharing site, 182

7.2 Social Technology Comes of Age

7.3 Blogging: The Internet Gets Personal

7.4 Social Networking

7.5 Social Bookmarking

7.6 Wikis

7.7 Media Sharing

Projects

Check with your instructor for the preferred method to submit completed project work.

Using Social Technologies in Business

Project 7A

Identify a company that is using some form of social technology, such as blogging, microblogging, wikis, social bookmarking, social networking, or media sharing. Determine the types of social websites that the company uses, its strategies in maintaining these sites, the use of these sites to reinforce the messages on the company's website, and the ability of these sites to attract interest from diverse groups of potential customers.

Project 7B TEAM

As a team, brainstorm a start-up business that you would like to launch. Write down the name of your business and the types of products or services that your company will offer. Establish your target audience and recognize your market competition. Discuss the ways in which social technology could be an effective communications and marketing tool to attract potential customers. Document the various steps that you would need to take to operate a business page on Facebook. Prepare a web-based presentation that introduces your business venture and outlines your Facebook business plan, including how you will drive people to your Facebook page and what kind of content you will place there to provide the information your customers need to interact with your business.

Gaining Experience with Blogging

Project 7C

To understand the purposes, types, benefits/drawbacks, and popularity of blogging, research a variety of blogs on the web. Find a blog that interests you and actively engage in a dialogue with other users by posting information and commentary. Prepare a memo that identifies the blog you chose and describes your experience with it. Include a transcript of your dialogue exchange in the memo.

Project 7D TEAM

Go to the Chapter 7 class blog. As a group, find a news story of interest and summarize the story on the class blog. Then have each team member respond to the summary by posting a blog entry once a day for a week. At the end of the week, prepare a transcript of the blog. Be prepared to discuss your exchange of ideas and opinions in class. Submit the transcript, as well as your source for the news article, to your instructor.

How Are Social Networking Sites Being Used Today?

Project 7E

Many colleges and universities today have their own social networking sites for their students. Go to your college's home page and locate the school's social networking site, if one is available. If your college does not have a social networking site, find a school that does have one. Where did you find the site? Is it current? Are there many participants? Does the site contain text, audio, photos, and/or videos? Prepare a summary of your findings.

Project 7F TEAM

As a team, investigate the impact of social networking on one aspect of our society, such as a political issue or a medical crisis. Determine the types of sites that were used, the topics that were discussed, the number of participants, and other ways social networking has influenced us today. Then, based on your research, predict the role that social networking might play in the future. Prepare a presentation of your team's findings and predictions.

Sharing through Media

Project 7G

Create a slideshow résumé. Post your presentation on https://ODW5 .ParadigmEducation.com/slideshare. Adjust your privacy settings so that your presentation can only be shared with your instructor and your class.

Project 7H TEAM

Using a video recording device, create a team video about one of the topics presented in this chapter. Upload your video to YouTube and share the link with your instructor.

Wiki Project—Building an Online Policy

Project 7I TEAM

You are employed at a large publishing company that produces newspapers and magazines. Recently, several of your publications have been posted online, and your company has created blogs on these sites. You are in charge of writing the blogging policies. Write a policy on the class wiki that addresses the following user issues:

- standards for appropriate behavior or content
- reporting of inappropriate behavior or content
- privacy safeguards

Before you begin writing, review the policies of mainstream blogging sites for ideas. When you have finished your blogging guidelines, submit the document to your instructor. Be sure to include any references you used in creating your document.

Project 7J TEAM

As a team, use the tools in PollDaddy or SurveyMonkey to create a survey on the use of social technology. The survey can be on any social technology topic, or you can use one of the suggested topics. For example, you may want to ask your survey participants how they are using personalized start pages or how blogging has benefited their businesses. Have each team member distribute the survey to 10 friends or family members. Share your survey results on the class wiki. When you post or edit content, make sure you include a notation within the page that includes your name or the team members' names and the date you posted or edited the content—for example, "Posted/edited by [your name or team members' names] on [date]."

Class Conversations

Topic 7A   What privacy issues does the social web create?

The social web is used by millions of people from many cultures and backgrounds. With so many people using these technologies, one issue that users face is privacy. What privacy protections do social networking sites offer? What personal information should be kept off social networks? What is meant by your social footprint? How could what you post on the web today have an impact on your future?

Topic 7B   How could you use social technologies to make a better world?

You work for a nonprofit organization working for a cleaner environment. How could you use social technologies to raise awareness and get new members and support? Discuss ways to use wikis, blogs, media sharing, microblogging, and social bookmarking in a campaign to encourage people to take public transportation or walk rather than drive a car.

Topic 7C   How should businesses use the social web?

People have jumped on the social web to connect with each other. Businesses are also actively using these tools to sell products and services to people by planting promotional comments in user pages. Should businesses be allowed to insert advertising into what was meant to be a way to hold a social dialogue? How do you feel about businesses using your social network profile to discover how to market products and services to you?

Digital Defense
Securing Your Data and Privacy

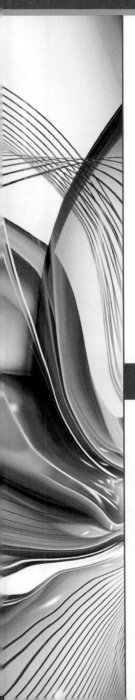

What You'll Accomplish

When you finish this chapter, you'll be able to:

8.1 Describe the risks associated with operating a computer connected to a network and the internet and list the tools you can use to protect your computing devices and data from those risks.

8.2 Explain the steps to secure a home network, the various types of personal computer and mobile device malware, and the methods used to obtain personal information from individuals.

8.3 Recognize security risks associated with mobile devices and with storing data in the cloud and give examples of tools and services to safeguard those devices and data.

8.4 Identify hardware and software tools and strategies used by organizations to secure corporate networks and prevent loss of data.

8.5 List security defenses that both organizations and individuals should adopt to prevent cyberattacks and data loss or theft.

Why Does It Matter ?

Would you leave your bank card and PIN sitting on an empty table in a food court at the mall? Would you leave your home for a vacation and not bother to lock the doors or windows? Many people who are used to protecting their wallets or houses may not take steps to guard their digital information against common threats. Even if you protect your computer with antivirus software, you might overlook routine tasks that can leave you vulnerable to losing important data. Everybody should learn the basic skills of computer security, because while replacing a computer is easy, replacing valuable data is not.

 The online course includes additional training and assessment resources.

Protecting the data on your computing devices and your personal information online are important concerns in our digital age. From firewalls and antivirus software to corporate security planning and protecting your mobile devices, learn about the tools available to help your information stay secure.

Course Content

Take a Survey

Technology in Your Future Video

8.1 The Role of Security and Privacy in Your Digital World

Do a search on a news website any day of the week and you'll come up with stories like these:

- In March 2018, Facebook made major headlines when it was reported that a loophole in Facebook's programming interface allowed data in 50 million user accounts to be misused. It was alleged that a company targeted ads designed to influence the outcome of the 2016 presidential election. Facebook was served with several lawsuits when its role was exposed and found itself fighting off a #DeleteFacebook movement.
- On September 7, 2017, Equifax (one of the three largest credit reporting agencies) admitted that a cybersecurity breach had affected about 148 million American consumers, as well as some Canadian and British citizens. A website vulnerability allowed criminals to access names, social security numbers, birthdates, home addresses, and some driver's license numbers. This was the largest data breach incident in 2017.
- In May 2017, hundreds of thousands of computers were attacked by the WannaCry ransomware, which spread rapidly around the world. WannaCry exploited a vulnerability in older Windows PCs that had not been updated to the latest operating system and rendered files on those PCs unusable, unless the owner paid a ransom fee. The attack was estimated to have spread to 150 countries with damages ranging from hundreds of millions to billions of dollars before it was stopped.
- Uber made public in November 2017 that hackers had stolen the data of 57 million Uber customers in 2016. Uber CEO Dara Khosrowshahi admitted that the company paid a $100,000 ransom to the hackers to cover up the data breach. Among many lawsuits filed against the company, the state of Pennsylvania launched a lawsuit against Uber in March 2018 for violating the state's data breach notification law.

The WannaCry ransomware attack took over users' computers, displaying messages such as this to extort payments.

Computer security and safety are very much in the news and on the minds of both company executives and individual computer users, but just what's involved in keeping your computing devices and data secure?

Computer Security: Where's the Threat?

Computer security, also referred to as *information security*, involves protecting the boundaries of your home or business network and individual computing devices from intruders. An important part of computer security is **data loss prevention (DLP)**, which involves minimizing the risk of loss or theft of data from within a network.

Security threats can originate from various sources. One of these sources is malware (such as computer viruses), which we will discuss later in the chapter. In other cases, employees' own negligence can cause a company to simply lose data, in which case these companies face the hard fact that their people have been their own worst enemy. Cyberattackers may damage or steal data and can come individually or in groups. Attacks may come from a malicious **hacker** or organized crime group, corporate spies, unethical employees, disgruntled colleagues, or, in the case of your home computer, from an ex-friend out to cyberbully you.

> ❝I think computer viruses should count as life. I think it says something about human nature that the only form of life we have created so far is purely destructive.❞
>
> —Stephen Hawking, physicist

Because of the wide variety of threats out there, it's important to recognize that proper security practices are not just a matter of individual protection—they are also critical for the overall security of the internet. If you allow your own PC to be co-opted by criminals to scam or spam others, you have become part of a bigger problem.

Figure 8.1 shows some of the measures you should put in place to keep your own digital world more secure.

What can happen if your information security is compromised? Companies can face enormous costs resulting from theft of their customer data and intellectual property. Kaspersky Labs, an internet security company, reported that the average cost of a data breach in North America in 2017 was $1.3 million. It's estimated that small- and medium-sized businesses have incurred an average cost of $117,000 for lost business and costs associated with hiring experts to help recover and restore IT infrastructure and data.

The Kaspersky report numbers do not include the massive breach at Equifax, which, at the end of 2017, had amassed costs of $439 million, with only $125 million recovered from insurance. Larry Ponemon, chairman of the Ponemon Institute (a research center for privacy, data protection, and information security policy) estimated the final cost to be nearly $600 million. Ponemon said this would make the Equifax debacle the "the most expensive data breach in history."

Cyberattacks on small businesses are increasing, according to research by the National Cyber Security Alliance. More than 70% of attacks target small businesses. As much as 60% of small- and medium-sized businesses that experienced a cyberattack went out of business within six months.

FIGURE 8.1 **The Secure PC**
Computer security is all about keeping data—from your company's recipe for soup to your own bank account information—safe from threats and loss.

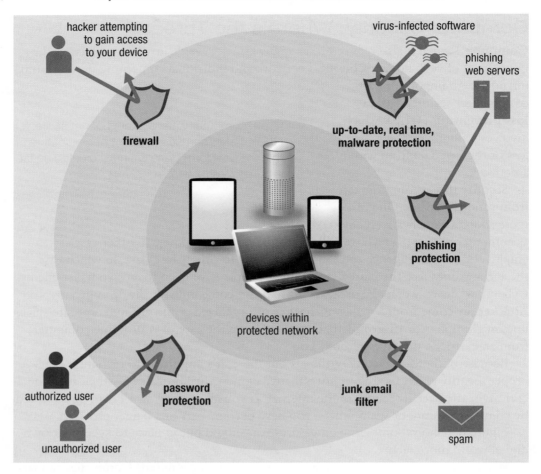

Of increasing concern to security experts is the **advanced persistent threat (APT) attack**. An APT is a type of attack where the cybercriminal gains access to a network for an extended period of time to steal data. A typical APT attack involves a hacker choosing a person or group at a target organization. The hacker then sends an infected email or other malware to gain access to the person or organization's network. In 2018, the US Department of Justice filed an indictment against an Iranian consulting firm that had been successfully using a form of APT attack since 2013 to steal academic research from universities, proprietary data from companies in the US, and government data.

For individuals, the primary security risk today is from **identity (ID) theft**, and the associated costs are tallied not only in dollars but in time as well. The Federal Trade Commission (FTC) tracked 2.7 million consumer complaints in 2017 with 14% related to ID theft, making ID theft the second biggest category of complaint. In the FTC report titled *Identity Theft: The Aftermath 2017*, 61.9% of ID theft victims reported that their cases were still not resolved after five years, and more than one quarter of those surveyed had to resort to borrowing money from family and friends to get by.

Basic Tools of Computer Security

High-profile hacks into companies such as Facebook and Equifax have companies reviewing, updating, and improving their security procedures. The tools you can use to protect your personal computer and information are similar to tools corporations use to protect their intellectual property. Both individuals and companies can implement authentication processes, security technologies, and user procedures to keep data safe.

Authentication involves the use of passwords and, in some cases, other identifiers such as fingerprints to make sure that the people accessing information are who they claim to be. For example, on your personal computer you can create a user name and password to ensure that others do not have access to your programs and data.

To protect online accounts, everyone should use sites that provide **two-factor authentication (2FA)**, also referred to as *two-step verification*, which adds a second layer of security. The most commonly used 2FA is typing your password and then receiving a code via text message that you enter on a second screen. Other types of 2FA require you to enter information you know, such as a PIN or the answer to a question, such as the color of your first car. The hacker is not likely to have the personal information about you or know your phone number as well as your password.

Security technologies that help keep out intruders or defend against dangerous computer code include firewalls, encryption, and antivirus software, among others.

User procedures may be as simple as teaching your kids not to follow a link online that might download a virus, or as robust as a company-wide policy identifying who can access data and establishing procedures for backing up files to avoid data loss. All computer users need to be suspicious of unsolicited links and follow their company's defined security practices at all times.

McAfee antivirus software is one of the products people use to protect their personal computers from damaging viruses.

The increasing number of households connected to the internet has resulted in an increased need for vigilance by individuals to defend against cyberattacks, identity theft, and other types of fraud. In addition, loss or damage to data resulting from inadvertently downloading damaging programs can be frustrating and costly. Historically, most security threats have targeted Windows users, but as Mac has gained double-digit market share in the last 10 years, and as Linux and Android operating systems become more prevalent, those systems have also become targets for cyber criminals. For example, in January 2018, Apple confirmed that Macs, iPhones, and iPads were all vulnerable due to flaws in Intel chips.

The prevalence of mobile devices in our daily lives has attracted hackers to the world of mobile malware. Originally, Android smartphones and tablets were the favorite targets for mobile malware, but recently Apple iPhones and iPads have fallen prey, too. According to McAfee, mobile malware caused 16 million infections in the third quarter of 2017—double the rate for the same period in the previous year.

> " Criminals want the biggest bang for their malware buck, which means the dominant operating systems, browsers, platforms, etc., are always going to be the better targets. "
>
> —Linda Criddle, privacy consultant at Intel Corporation

Tony Anscombe, a security analyst for another popular antivirus company, AVG Technologies, warns that hackers will take a popular app, such as Candy Crush Saga, insert malicious code into the program, and then publish the app on a third-party site. This allows them to avoid the malware detection mechanisms found on Google Play or the Apple store. Bluetooth users must also beware of mobile security threats. Although Bluetooth connections are short-range, they can be used to intercept data or to send harmful files or viruses.

Protecting Your Home Network

If you have set up a home network that enables your computers to connect to the internet and you haven't thought about security, it's as if you just installed a back door to your house that's left wide open 24/7. An unprotected network means that anybody who is near your home can "piggyback" on your internet connection, track your online activities, and possibly even hack into your computer.

You can take several simple steps to secure a home network. Wi-Fi home networks use an access point or router, which is a piece of equipment that comes with a preset password. The bad news is that these default passwords are pitifully predictable and simple. The good news is that you can go to the device manufacturer's website and find instructions for changing the password to something that is harder to guess.

Another important step in securing a home network is to use **encryption**, a part of cryptography, which is the study of creating algorithms and codes to protect data. Encryption scrambles a message so that it's unreadable to anybody who doesn't have

Routers have some built-in protections for your network, if you know how to use them.

the right key. Say you want your friend to send you a message that will contain data you want to protect. You use your computer to generate a **public key**, which you send to your friend. Your friend applies the key, which encrypts the message, and sends the message to you. Your computer then applies a **private key**, known only to you, to decrypt the message. Because the message is encrypted, nobody but the intended recipient can read it. Figure 8.2 depicts this example of the **public key encryption** process. Two forms of encryption are **Wi-Fi Protected Access (WPA)** and **Wired Equivalent Privacy (WEP)**. **Wi-Fi Protected Access 2 (WPA2)** uses a stronger, more complex form of encryption than WPA.

In 2018, the security standard **Wi-Fi Protected Access 3 (WPA3)** was finalized. WPA3 will appear in new devices late in 2019. It includes features to make it harder for hackers to gain access to your wireless network with password-cracking tools, improves the security of public Wi-Fi networks, and makes it easier to connect IoT devices to the Wi-Fi router in your home. Check that your router uses the strongest encryption—WPA2 or WPA if WPA3 is not available.

The Menace of Malware

Collectively, nasty computer programs such as viruses and spyware are called **malware** (*mal* means bad or evil in Latin and *ware* refers to software). Malware installs itself on your computer without your knowledge or consent. Malware can do anything from pelting you with pop-up window advertisements to destroying your data to tracking your online activities with an eye toward stealing your identity or money.

In the early days of computers, individual hackers often planted viruses just to aggravate people or exploit a technological weakness. Today, most malware is created by less-than-ethical businesses, organized gangs, or criminals who aim to download dangerous code to your computer, co-opt your email contact list to send out **spam** (mass emails), or perform other illegal activities for profit-based motives. The following are descriptions of some common forms of malware.

Viruses   A **virus** is a type of computer program that can reproduce itself by attaching to another, seemingly innocent, file. Viruses duplicate when the user runs an infected program. A typical scenario is that a virus is part of an email attachment. This

FIGURE 8.2   **Public Key Encryption**
The key for encrypted data is like your house key—it unlocks the data.

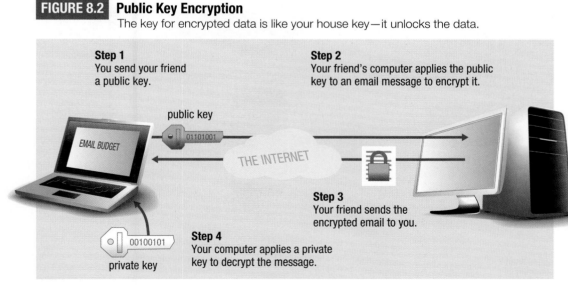

Step 1
You send your friend a public key.

Step 2
Your friend's computer applies the public key to an email message to encrypt it.

public key
01101001
EMAIL BUDGET
THE INTERNET

Step 3
Your friend sends the encrypted email to you.

Step 4
Your computer applies a private key to decrypt the message.

00100101
private key

is one of the ways viruses spread from computer to computer. When the user opens the attachment, the program runs and the virus is duplicated. If the user does not open the attachment, the program does not run and the virus does not duplicate itself. Many viruses eat through your data, damaging or destroying files. Figure 8.3 illustrates the ways in which a virus attacks.

Worms A **worm** is also a self-replicating computer program, but it doesn't have to be attached to another file to do its work. A worm does not require the user to do anything. If your computer is connected to an infected network, you can put it at risk merely by powering it on. A worm has the nasty ability to use a network to send out copies of itself to every connected computer. Worms are usually designed to damage the network, in many cases by simply clogging up the bandwidth and slowing its performance. Figure 8.4 shows how a worm attacks.

Trojans Named after the infamous Trojan horse of Greek legend, a **Trojan horse** is malware that masquerades as a useful program. When you run the program, you let the Trojan into your system. Trojans open a "back door" to your system for malicious hackers, just as the Trojan horse allowed invaders to enter a city and then attack from within. Trojans are becoming more sophisticated, often disguising themselves as authentic operating system or antivirus warning messages that, when clicked, download the Trojan malware to your computer or mobile device. Figure 8.5 shows how a Trojan horse attacks.

FIGURE 8.3 **How a Virus Attacks**
When you forward an email with an attachment such as a picture, you may be spreading a damaging virus.

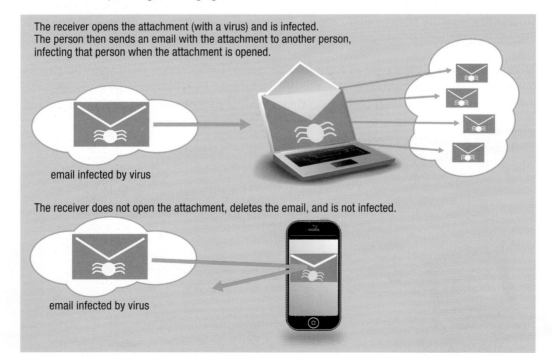

The receiver opens the attachment (with a virus) and is infected.
The person then sends an email with the attachment to another person, infecting that person when the attachment is opened.

email infected by virus

The receiver does not open the attachment, deletes the email, and is not infected.

email infected by virus

Our Digital World

FIGURE 8.4 **How a Worm Attacks**
A worm reproduces itself and attacks all the computers on a network.

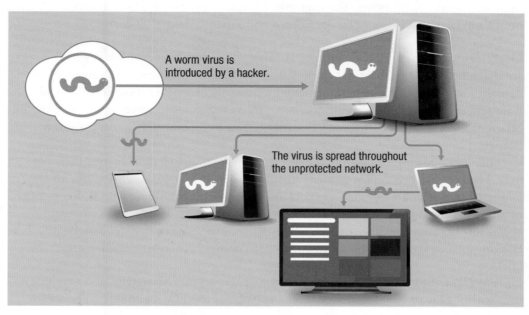

FIGURE 8.5 **How a Trojan Horse Attacks**
A Trojan horse pretends to be a useful program but ends up opening your system to hackers.

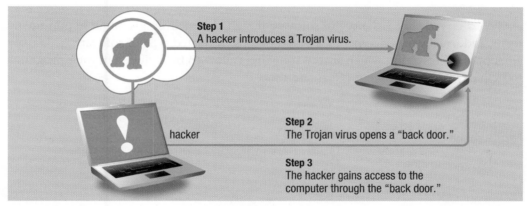

Macro Viruses and Logic Bombs Other malicious programs come in the form of small pieces of code embedded in a program. A **macro virus** is usually found in files such as word processing documents and spreadsheets and can corrupt the computer when the user opens the document and executes the macro (a recorded series of keystrokes that can be played back to perform a task). A **logic bomb virus** might be placed in a software system to set off a series of damaging events if certain conditions are met (for example, if you try to delete a set of files).

Rootkit A **rootkit** is a set of programs or utilities designed to gain access to the "root" of a computer system or the system software that controls the hardware and software. With this access, a hacker can then monitor the user's actions. This can take place on an individual system or on a network system. An important aspect of a rootkit is that it cannot be detected, at least not easily, by the user (or the administrator, in the case of a network). While rootkits can serve harmful purposes, they can also be used for legitimate purposes. For example, programs used by parents to monitor children's internet activities can be considered rootkits.

Botnet A **botnet** is a collection of **zombie** (or robotlike) computers, which are machines that have been taken over by malware for the purpose of causing denial-of-service attacks, generating spam, or conducting other mischief. The malware sets up a stealth communication connection to a remote server controlled by the cybercriminal. Some security experts attribute most criminal activity on the internet to botnets.

There's a growing trend of hackers co-opting smaller IoT devices into their botnets. Hackers used to deploy botnets mainly against Windows-based PCs. With the increased use of smart devices in offices and homes, this trend has caused the number of zombie attacks to increase substantially. Security experts have raised concerns about this because IoT devices typically don't receive security updates as regularly as a Windows PC. Figure 8.6 shows how malware creates botnets to organize an attack.

Spyware **Spyware** is aptly named because it spies on the activity of a computer user without his or her knowledge. Some spyware is used by legitimate websites to track your browsing habits in order to better target advertisements to you. Spyware can also be used by businesses to track employee activities online. However, other spyware, such as a keystroke logging program, can be used by criminals to learn your bank account number, passwords, social security number, and more.

Adware **Adware** is a piece of software designed to deliver ads, often in pop-up form, and usually unwelcome, to users' desktops. A related type of software is ad-supported software, which shareware writers allow to be included in their programs to help pay for development effort and time.

FIGURE 8.6 **How Malware Uses Botnets**
Your computer can be taken over by bots and used to send spam or malware to others.

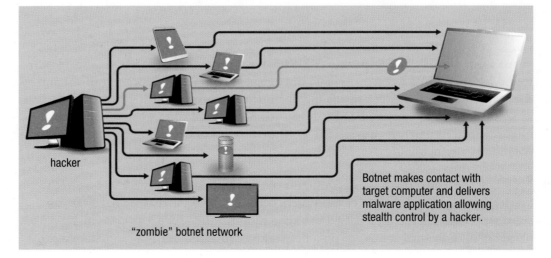

hacker

"zombie" botnet network

Botnet makes contact with target computer and delivers malware application allowing stealth control by a hacker.

Products such as Malwarebytes' Anti-Malware scan your computer for malware; some other products help prevent it from downloading in the first place.

Scareware Scareware is a scam in which an online warning or pop-up convinces a user that his or her computer is infected with malware or has another problem that can be fixed by purchasing and downloading software. In reality, the downloaded software may not be functional or may itself be malware. These scams are primarily used to steal the user's money and credit card information.

Ransomware Ransomware is a scam in which access to one's computer is locked or restricted in some way. In some cases the contents of the hard drive are encrypted. A message displays to the owner of the PC or mobile device demanding payment to the malware creator to remove the restriction and/or restore the owner's data. Ransomware is growing. In late 2017, a cybersecurity research and market intelligence firm predicted that a ransomware attack would occur in a business every 14 seconds by the end of 2019. The cost of the damage inflicted by such attacks could reach $11.5 billion annually.

How Is Malware Spread?

There are several ways in which malware, depending on its nature, can be spread:

- You can infect your computer by tapping or clicking an email attachment that contains an executable file.
- Pictures you download can carry viruses stored in a single pixel of the image.
- Visiting an infected website can spread malware.
- Viruses can spread from a computer storage device such as a DVD or flash drive that you use on an infected computer and then insert into another computer drive.
- Worms can spread by simply connecting your computer to an infected network.
- Mobile devices can be infected by downloading an app, ringtone, game, or theme that carries malware.
- A mobile device with Bluetooth enabled in "discoverable mode" could be infected simply by coming within range of another Bluetooth device that has been infected and is running the same operating system.

Security threats are a reality in our digital world. What's also true is that several programs and technical tools are available to protect your computer against these potential hazards, as explained later in the chapter. In addition, knowing how to recognize trustworthy websites and how to manage cookies are two proactive strategies everyone can use.

> **Playing It Safe**
>
> Be especially cautious when you receive a chain letter via email. These are often simply devices for delivering malware or collecting email addresses for the purpose of building spam lists.

Look for these symbols from various organizations that verify the secure practices of sites.

Recognizing Secure Sites

Although even a reputable site may occasionally pass on a dangerous download to your computer, it's the sites that actively download malware that you have to be most cautious about.

Buying only from reputable businesses that sport various accreditations such as those from TRUSTe, SiteTrust, and ValidatedSite is one step toward safety.

Sites where you perform financial transactions should always have **Transport Layer Security (TLS)** in place. Developed from an older cryptographic protocol called **Secure Socket Layer (SSL)**, TLS is a protocol that protects data such as credit card numbers as the data is transmitted between a customer and online vendor or payment company. In a web browser, two things can signal that you are using TLS: "http" in the address line is replaced with "https," and a small closed padlock appears next to the address bar or in the status bar of the window.

Other useful tools are products such as McAfee Site Advisor or similar tools built into browsers and security programs. These display an icon next to sites in your search results indicating websites that are known to have doubtful business practices or to routinely pass on malware to visitors.

It may be safer to do business with retailers you know from the "brick-and-mortar" world. Also, you should always type a URL into your browser to go to a site rather than tapping or clicking a link in an email or advertisement.

Managing Cookies (Hold the Milk)

A **cookie** is a file stored on your computer by a web server to track information about you and your activities. Cookies can be completely harmless and even helpful. For example, if you shop at an online store often, when you next visit that site you might find that the store knows your name and has suggestions of items that might interest you. Sites can provide this personalized service by reading the information stored in the cookie.

However, some companies or individuals plant cookies on your computer for other reasons. They may be trying to track your activities to gather enough information to steal your identity, for example.

Every major browser has tools and settings for dealing with cookies. For example, in Microsoft Edge (shown here), you can adjust the settings to accept all cookies, block cookies from certain sources, or block all cookies.

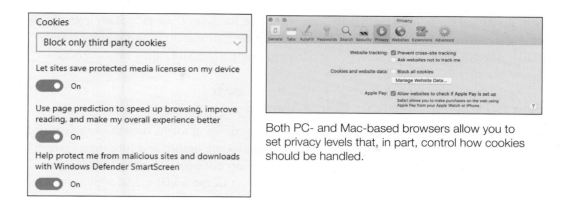

Both PC- and Mac-based browsers allow you to set privacy levels that, in part, control how cookies should be handled.

Foiling Phishers: No Catch Today

Phishing (pronounced "fishing") refers to the practice of sending email that appears to be from a legitimate organization in an attempt to scam the user into revealing information. Typically, the revealed information is used for identity theft. The email directs the user to tap or click a link that then goes to a bogus website that appears valid, often containing logos and color schemes that simulate the real organization's branding. The bogus site prompts the user to update personal information such as the user name, password, credit card number, or bank account number. Once you follow these links and enter your information, you've basically handed over your sensitive information to criminals.

Delete any messages you receive that ask you to update information in a financial or retail account. Never follow a link to bank or other financial websites—always enter the URL into the browser address field yourself.

Advances in filtering technologies used by email providers and increased awareness of phishing attempts by the general public have forced hackers to innovate. Increasingly, hackers are directing their phishing techniques to customers of cloud providers. Rather than the typical financial or credit card request, hackers are seeking valid sign-in credentials to gain full access to cloud data. For example, in the summer of 2017, Gmail users received an email that appeared to be legitimate. It directed the recipient to an actual Google page. Individuals were asked to grant access permissions to a malicious third-party program planted by the hacker. Once an individual complied, the hacker was able to view the person's contacts, emails, and even see their location and files. Phishers are targeting cloud providers to mass harvest sign-in credentials (such as for Gmail accounts) because individuals often use the same email address and password combination for multiple online websites. This practice exponentially increases the payoff for the hacker.

The SANS Institute (an information security and research organization) reported that 95% of all attacks on enterprise networks are the result of successful **spear-phishing** campaigns. In a spear-phishing attack, a person is targeted because of where he or she works. The hacker sends the individual a phishing email to get that person to reveal a trade secret or other data that lets the criminal gain financially. A digital security firm predicted early in 2018 that hackers would be focusing their tactics on lower level employees who have access to sensitive company data (researchers and administrative staff). Aryeh Goretsky, a researcher at antivirus company ESET, aptly says, "*Think before you click.*"

Social Media Risks: Click Wisely

Social media malware is becoming one of the most common forms of malware infection. Scammers use programs in social networks to impersonate brands and lure their customers to the cybercriminal's door. Antivirus programs do not protect users of social networking sites because the malware is operating as part of the social media application.

Strategies used by cybercriminals include:

- **Phishing a family member.** In the spring of 2017, a phishing attack that was launched on Twitter successfully netted an employee of the US Department of Defense. The employee's spouse was the source of the breach when she clicked a link to a vacation package after talking online about the family's summer holiday plans.

- **Profile cloning.** In this technique the hacker impersonates an individual's account and interacts with a target, making it more likely that the unsuspecting individual will share information with the imposter or click a link that leads to a malicious site.

- **Impersonating a brand or manipulating content.** Hackers can create fake accounts designed to impersonate a company or a brand's support site. They can then spread malware to customers who are lured to the fake pages. Hackers also use these fake accounts to spread content using clicks and shares that end up affecting the news feed. Another frequent strategy is using fake accounts or botnets on social networks for click-fraud scams.

- **Invitations to games.** To play some games online, you may have to grant access to your profile, friends list, email address, and your birthday. Malicious hackers will then use the information to steal your identity, apply for a credit card or loan in your name, or transfer money out of your bank account.

> **🔒 Playing It Safe**
>
> Remember, links aren't the only online trap you need to avoid. Viruses can be contained within pixels of pictures or in files that you download. If you forward any kind of message to a large group of people, their email addresses could be captured and used for ID theft or other purposes. You may be told in a message that you've won a contest, but before you respond, remember, if you didn't enter a contest, you couldn't have won a prize.

Hackers are succeeding on social networks because individuals are more comfortable clicking links on sites like Facebook and Twitter than they are in their email inboxes. Victims believe they are safer because they have chosen to follow their friends and the companies with which they like to do business. Individuals are also more prone to share personal information on their favorite social network—opening the door for a hacker's exploits.

Mitigating your risk on social networks includes taking a moment to think before clicking a link in a message.

- Ask yourself if the message makes sense and if there are spelling and grammar errors or a tone in the content that alerts you that the message is not from your friend.

- Change your passwords frequently on your social networks and avoid using the same password for all of them.

- Always log out of the network's site after checking in—especially on your smartphone.
- Lastly, avoid clicking links to ads making preposterous claims, such as those claiming that a celebrity has done something outrageous or has passed away.

> **◖ Course Content**
> Take the Next Step Activity
> Ethics and Technology Blog

8.3 Mobile Security

Computing is no longer just a sit-at-your-desk activity—it's mobile! Mobile computing is convenient, but it also brings its own security risks. Various settings and tools can help you keep your portable device and the information stored on it safer.

Protecting Your Laptop or Tablet and Its Precious Data

When you bring your computer with you, you're carrying a big investment in both dollars and data, so it's important that you protect it from theft and damage. This is a concern for both personal users and corporations, who are struggling with protecting IT assets as their workforces begin to carry smaller devices that can be more prone to being lost or stolen. Consider using one or more of the following devices and procedures to help you secure your laptop or tablet computer.

Locks When you ride your bike downtown, you probably secure it with a bike lock once you reach your destination. Laptops can be secured using comparable devices that tie them to an airport chair or a desk in a field office to deter potential thieves from snatching them. As with bike locks, the determined thief with enough time can cut this cable and get away with the goods, so it's only a slight deterrent.

Remote Tracking/Wiping To protect your data in the event of a theft, consider services such as LoJack for Laptops, which allows you to remotely delete data and use GPS to track your wayward laptop or tablet. More than one Android app is available that enables you to track and lock your Android tablet remotely, and the Apple's Find My iPhone app can be used to locate your iPhone, iPad, iPod touch, Mac, Apple Watch, or AirPods. Expect remote security options to increase as tablet technology evolves. Remote tracking and wiping is not just for laptops and tablets—makers of smartphones provide similar services to track and wipe lost or stolen phones.

Fingerprint Readers Many newer laptops include fingerprint readers. Because fingerprints are unique to each individual, being able to authenticate yourself with your own set of prints to gain access to your computer is a convenient and effective security feature. If somebody without a fingerprint match tries to get into the computer data, the system locks up.

Traditional cable locks allow you to physically tie down a laptop. Fingerprint readers restrict access by matching authorized fingerprints.

Password Protection If you travel with a mobile computer, activating password protection and creating a secure password or passcode is your first line of defense. If somebody steals your laptop or other mobile device and can't get past the password feature, he or she can't immediately get at your valuable data. Activate the passcode/lock feature on your tablet and configure the device to lock automatically if it is inactive for a set period of time. This action could help protect your data if you lose your device, since there is a time lag from when you initially become aware of the loss of your device and when you take action to protect or recover your data. It is during this delayed response time that your data is particularly vulnerable.

Mobile Computing Policies for Employees Stopping thieves is one concern when you're on the road, but stopping employees from making costly mistakes regarding company data is another area where companies must take precautions. Making sure that employees who take company laptops or tablets outside of the office are storing these devices safely and securely offsite is vital to company security. Policies might require them to keep backups of data on physical storage media such as a flash drive, or to back up data to a company network.

Using Wi-Fi Safely

When you access the internet using a public hotspot (a location that offers Wi-Fi access), you have to be very careful not to expose private information. Anything you send over a public network can be accessed by malicious hackers and cybercriminals. Limit your use of online accounts to times when it's essential. Be especially on guard when accessing your bank accounts, investment accounts, and retail accounts that store your credit card for purchases, and avoid entering your social security number.

Public internet access is incredibly convenient, but in order for businesses such as internet cafés to provide access to the public, they have to remove vital security settings.

One type of attack that occurs in public areas with free Wi-Fi hotspots is known as a **man-in-the-middle (MITM) attack**. Public Wi-Fi routers are often unsecured or poorly secured, providing a hacker with easy access to the router's configuration. The hacker can then set up tools that read the data being sent from one device to another. A typical scenario is when a person uses public Wi-Fi to sign into a personal email account or corporate network. The hacker grabs the credentials and then forwards the information to the intended destination unaltered so that the sender has no idea that her data has been captured by a cybercriminal.

Mobile Phone Security

There are three major security issues when it comes to using smartphones, which are the most common mobile computing devices in use today. First, you have to protect the phone from loss or theft. Second, you should be cautious if your phone features Bluetooth. Third, you need to avoid mobile viruses.

- If you don't want to pay for a third-party app, all of the smartphone OS providers include free find-and-wipe-clean utilities for lost and stolen phones. These include Android's Find My Device, Apple's Find My iPhone, and Microsoft's Find my phone for Windows phones. You may have stored contact information as well as some passwords to access online sites on your phone, and these are at risk if your phone is misplaced or stolen. What can you do to protect this sensitive data? Some services have started to provide data protection for mobile phones. For example, when your phone is lost or stolen, you call the service, and it clears data from the SIM card (the card that holds all phone data such as your contacts), causing the phone to emit a shrill sound and the keypad to lock. One company is developing software that can track your speech patterns and walking gait—if it can tell somebody else is using your cell phone, it can lock up and require a password to proceed.

- Bluetooth technology allows devices in close proximity to communicate with each other—for example, a wireless printer might connect with a computer, or a smartphone might connect with a car Bluetooth system to make hands-free calls. However, once you turn on Bluetooth, any other Bluetooth-enabled device near your location can connect with your device.

    Two threats unique to Bluetooth are bluejacking and bluesnarfing. **Bluesnarfing** occurs when a cybercriminal gains access to your smartphone via your Bluetooth connection, intending to steal all the personal data and files on your phone. In a **bluejacking** attack, a hacker uses your Bluetooth connection to send information to another Bluetooth device. In a bluejacking attack, while

When you're on the go, you may be exposed to dangers and too rushed to take precautions, which could be a costly mistake.

the connectivity is compromised, your personal data may not be accessed. How can you protect your phone if you use Bluetooth? It's important to turn off Bluetooth when you aren't using it to avoid connection by nefarious people lurking nearby. If you don't want to turn off Bluetooth, look for a setting that allows you to hide the device so others can't discover it. Finally, don't accept a pairing request from a device that you don't recognize or with which you did not initiate contact.

- Mobile malware that can attack smartphone operating systems is becoming increasingly prevalent. Malware writers prefer Android because it is an open platform and has the largest market share; however, other mobile devices are also at risk of malware. To protect your phone, install and keep updated mobile security software on your device. Apply the latest updates for your mobile operating system as they become available. Finally, be cautious when downloading new apps to your device, because you could be inadvertently downloading malware. Avoid third-party app stores, and be mindful of where you are tapping or clicking while browsing the web on your device.

The popularity of BYOD (bring your own device) policies in the workplace has brought new security risks to the attention of company owners and managers. Allowing an employee to use his or her personal smartphone sounds like a money saver, but it also comes at some risk. One in three mobile devices are lost or stolen, which could expose company data stored on the lost device. Mobile malware an employee may have inadvertently installed can also put corporate data at risk by opening a back door into the corporate network.

Statistics show that less than half of smartphone users set up a password or passcode for their device. For a corporation that allows a BYOD policy, there should be a requirement for users to protect their devices with passwords. They should encourage users with iPhones or iPads to use Apple's Touch ID feature (iPhone 5s and iPad fifth generation or later), which unlocks the device using the owner's fingerprint. The iPhone X has Face ID technology, allowing users to unlock the phone with light scans that identify their face. Two other examples of this technology are Samsung Pass and Windows Hello, which allow you to unlock a device with a fingerprint, face, or iris scan.

Keeping a close watch on your phone and writing down any hardware or SIM card codes you might need to track or identify it are good smartphone safety practices.

With smartphones such as Apple's iPhone X, facial recognition is used to unlock your device—the biometric authentication is called *Face ID*.

Security in the Internet of Things

Do you have a smart thermostat, smart coffee maker, smart speaker, or smart TV in your home? Gartner, Inc. forecasts that 20.8 billion connected things will exist worldwide by 2020. While these smart devices are designed to make our lives easier, there are risks to having so many common household items connected to the internet. Consider these events that have already occurred:

- A smart fridge was hijacked to send spam.
- A baby monitor was hacked to spy on a toddler.
- Thousands of security cameras were hacked and then used to overwhelm a web server to the point where the website crashed.

The security concerns for IoT devices involve devices being taken over by cybercriminals intent on stealing data, causing harm to individuals, or adding to their botnet network. IoT devices may not be set up with regular software updates that close loopholes against new vulnerabilities.

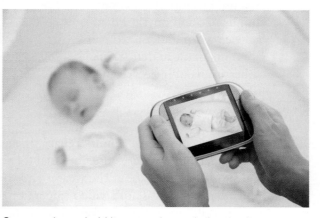

IoT developers have rated security as a top priority. This emerging field will continually update and evolve as developers look for ways to make IoT devices more secure and as

Common household items such as a baby monitor can pose a security risk if they are smart devices connected to your home Wi-Fi network.

hackers look for ways to thwart these attempts. New threats are created all the time. If you have a smart device connected to your home network, learn about the device's security settings, how the software is updated, and steps you can take to make sure your home network is secure.

Data Privacy in the Cloud

If you store your data in the cloud using a service such as Microsoft's OneDrive or Dropbox, learn about the provider's privacy and security policies with regard to your data. These sites typically allow you to share information only with specific users and to set other policies and permissions. However, there may be other ways for unauthorized users to gain access to your data, such as employees at the host site being able to view the contents of your files. Protect yourself by considering the sensitivity of the data before you post it. For example, don't post files such as your tax return, which includes your address and social security number, because the exposure of that data would make you vulnerable to identity theft.

Securing data in the cloud is also a concern for business owners and managers. In July 2017, World Wrestling Entertainment exposed the data of 3 million fans on an Amazon Web Services server. Data was erroneously set to "Public," making customer information (including names, email addresses, postal addresses, age, race, and gender) available for anyone to download. This is just one example of several data breaches that occurred during the year. The rapid adoption of cloud-based applications has caused growing pains for businesses that use them without proper security training and procedures.

Data Privacy and Wearables

Wearable technologies such as fitness trackers and smartwatches are popular, and they do a lot more than count your steps or record your heart rate. Other wearable technologies include:

- a video camera that resembles a button on your shirt or blouse
- a medication with an ingestible digital sensor (called a *digital pill*) used to monitor patient compliance and physical data
- a device worn under a woman's garments that can detect the earliest signs of invasive cancer in breast tissue
- a smart swim suit that tracks your pool laps

Data transmitted from a smartwatch is vulnerable to interception.

As wearables become more commonplace, the amount of data being stored in the cloud and on smartphones raises the potential for privacy and security concerns. These devices collect information about the wearer and the physical health and wellbeing of that individual. In the case of a video camera on a shirt or blouse, the camera can record other people around the wearer without their knowledge. This represents a treasure trove of information that can be mined, aggregated, and used for purposes not originally intended by the owner.

Consider other people who have connections to the wearer who might be interested in the data that has been collected—an insurance company, an employer, a competitor—one can't be sure that these other people don't gain access to personal information.

Symantec, a security software company, outlined three risk points for data collected from a wearable device:

- On the tracker or smartphone where the data is saved, a hacker could use malware to access it.
- During transmission from the device to and from the smartphone and during transmission from the smartphone to and from the cloud, data could be intercepted by a hacker as it passes through Bluetooth or Wi-Fi networks.
- While stored in the cloud, the data could be breached by a hacker or exposed by human error.

Like the IoT, security and privacy development around wearable technology is an evolving field. Users should stay alert to new information about these devices, read privacy policies, and not blindly accept terms and conditions when they set up accounts. Devices should be turned off when not in use. You should always protect your smartphones and cloud accounts with a strong PIN or password.

◑ Course Content
Take the Next Step Activity

8.4 Security at Work

Loss of your personal work, such as a report to a club or organization you belong to, might be annoying, but loss of your company's customer list or secret recipe could be devastating. For that reason, many corporations have sophisticated security strategies and tools in place to protect them against outside attacks and data loss due to negligence by employees.

Corporate Security Tech Tools

Organizations take advantage of some security tools that most individuals don't need. For example, though encryption is available to the average person, few encrypt the majority of their data and messages. However, most companies use some form of encryption on a regular basis.

Symmetric encryption is often used by companies to ensure that data sent across a network is kept secure from outsiders. Companies can protect stored data and data in transit on the internet or across a corporate network by designating which computers should be able to share encrypted data and then providing them with the matching keys. The computers use the key to encrypt and decrypt data sent among them. Figure 8.7 shows the process of symmetric encryption in action.

Two other tools used by businesses that conduct online transactions are digital signatures and digital certificates. A **digital signature** is a unique string of characters used

FIGURE 8.7	**Symmetric Encryption**

Only devices that have the key that matches the one used to encrypt the message can decrypt the message. The encrypted message can be shared within a network or between two networks.

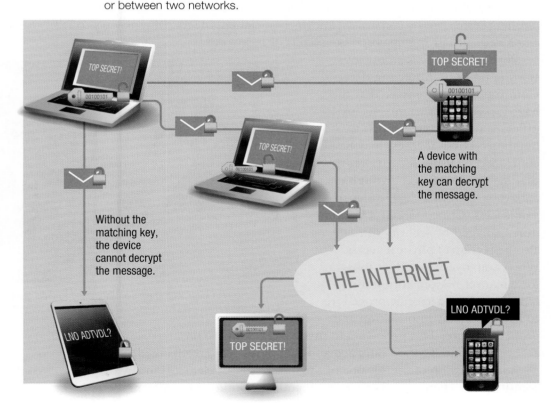

to verify the authenticity of a file that travels over the internet. The unique characters rely on public key and private key encryption to work. A **digital certificate** takes a digital signature one step further. A business with a digital certificate has been authenticated by a third party that validates the owner and issues the certificate. When you sign on to a secure site, your browser verifies the signed certificate so that you know that the certificate belongs to the company you are working with. If you receive a message in your browser that a certificate is invalid or out of date, the safest action is to note the error message, cancel the transaction, and notify the company.

An **Intrusion Prevention System (IPS)** is a robust form of anti-malware program that offers network administrators a set of tools for controlling the system. An IPS detects malware and can block it from entering the corporate network. One type of IPS can also detect suspicious content or unexpected traffic; this is called an **anomaly based intrusion detection system**.

Many corporations also use tools to audit events on their networks. These audits help to spot random problems or attacks. One technique is to create a so-called **honeypot**, which is a computer set up to be easily hacked into. When an attacker targets the honeypot because it's the easiest prey, companies can identify weak spots in their security.

Preventing Unauthorized Access

Individuals or groups who might target a company's data range from criminals trying to steal corporate secrets or customer credit card numbers to disgruntled employees looking for revenge. Preventing unauthorized access to the most sensitive data may involve both physical security (for example, locking the door to the network server room) and an electronic authentication system that requires users to identify themselves. It is also important to provide employee training to help employees avoid phishing scams run by con artists, known as **social engineers**, who may try to talk employees into giving up corporate secrets or passwords.

Physical security is an obvious starting point in protecting data. Companies should ensure that they have locked server rooms, secured offices, and controlled access to buildings. Physical access may be controlled by using security cards that have to be swiped through or passed in front of a card reader to gain access. Closed-circuit TV monitors managed by security officers also help control physical access. When employees lose their jobs for whatever reason, corporations typically follow a specific procedure to keep company information and

Anyone who has ever lost their security card knows that getting back into the office can be a challenge—which is exactly the point!

property secure. This policy might involve checklists to ensure that all employee access cards and keys are returned and that passwords are changed.

Authentication of users is one of the most essential elements in any corporate security plan. There are several levels of authentication, ranging from the input of a simple user name and password to the use of **biometrics**, which involves using devices such as a fingerprint reader or face or voice authenticator to identify individuals by a unique physical characteristic.

Criminals constantly try to find ways around authentication systems that require simple user names and passwords. For example, **spoofing** is a technique used by malicious hackers to make it appear that they are someone else so that they can convince a user to give up valuable information. Corporations have become more and more aware of such social engineering attacks. For example, a

Our fingerprints are unique, so a fingerprint scan is a good way to prove identity.

criminal might call into the company office at midnight, tell a security officer that he is an employee working on the road, providing enough personal information gained by various means to convince the officer. The story usually goes that while on a business trip he lost his access information to get into the network where vital files reside for his meeting the next morning. The officer goes to the requested office, locates secure information, and provides access to the network. The crook is in and can then have a field day with company data.

Denial-of-Service Attacks

A **denial-of-service (DoS) attack** targets a corporate system with continuous service requests so constant that response time on the system slows down and legitimate users are "denied service." A DoS attack typically causes an internet site to become inefficient or to completely crash (fail). Targets of these attacks are quite often high-profile sites such as banks or internet service providers such as AOL. Attacks may involve a set of distributed computers all pumping out requests to the target system until the system slows or fails. A DoS attack can result in slow performance of a site, unavailability of a site, or a huge number of spam emails being received. Figure 8.8 shows how a DoS attack might occur.

What are the motives for DoS attacks? Some are launched by spiteful individuals or groups. Others may be terrorist attacks or an assault by a competitor to damage the competing business.

Whatever the motive, companies typically use a three-pronged approach to combat DoS attacks:

- Harden the network against attacks.
- Detect intrusions.
- Block malicious actions.

Many software tools are available to address the three parts of this strategy, and the security industry is continually developing new programs to counteract the latest DoS threats. Network administrators often use a combination of firewalls, antivirus products, and the like to prevent attacks. Detection software looks for behavior patterns and characteristics of known attack types. Ideally, products recognize the content of network traffic quickly enough to block attacks.

FIGURE 8.8 | **Denial-of-Service Attack**

Denial-of-service attacks might target a government or company network.

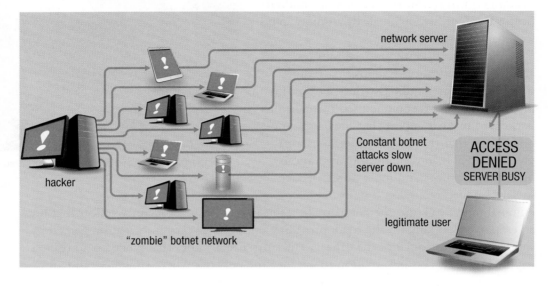

Data Protection in Disastrous Times

Just as individuals have a first aid kit and some extra water and a flashlight around in case of disaster, companies typically have a data **disaster recovery plan (DRP)** in place. Damage of records, from paper to electronic files, can happen on a large scale when an earthquake or hurricane hits or when an unauthorized user gains access to the company network.

Companies typically set up their networks to back up regularly, which may be daily or more often so all information stored on their computers is kept safe. Individuals may also use automated backup features that run at regular intervals, or they may back up manually to a storage medium such as a DVD or flash drive.

To back up company servers that contain a great deal of vital company data, there are three options:

- A **cold server** is simply a spare server that can take over server functions.
- A **warm server** is activated periodically to get backup files from the main server.
- A **hot server** receives frequent updates and is available to take over if the server it is connected to fails. (The process of redirecting users to this spare server is called **failover**.)

Many large corporations use an off-site backup. For example, a television or movie company might create copies of their programming and store them at another facility in case an original is lost or damaged or there is a natural disaster.

Disasters that can destroy important company data come in all shapes and sizes and are best prepared for by backing up data.

To avoid loss of data from a sudden surge in power such as might occur during a thunderstorm, individual computers can be plugged into a **surge protector**. In addition, an **uninterruptible power supply (UPS)** can provide a battery backup that takes over in the event of a power failure. UPS systems typically provide backup for about 15 minutes and require an auxiliary power supply to kick in during that time.

Employee Training

The final piece of corporate security is to train employees in the procedures that will keep corporate intellectual property and other data safe. Employees are more likely to comply with security policies if they understand why they are needed.

Employees should be trained in security measures such as using strong passwords and changing them frequently; using a swipe card to access the company premises; and keeping certain company information, such as network access codes, confidential. Many IT departments offer employees regular security training to educate them about the latest cyberattack methods and best practices for keeping data secure.

Computers in Your Career

Cyberforensics, also known as *computer forensics* or *digital forensic science*, is a field that requires special skills and training. If you worked in this field, you would spend your day extracting information from computer storage that can be used to provide evidence in criminal investigations or identify terrorists. This might involve decrypting data or finding residual data on a hard drive that somebody has tried to wipe clean. **Mobile forensics** relates to finding data saved or sent via a mobile device. Several colleges offer majors in the field, so if you like to solve mysteries, consider cyberforensics as a career.

Course Content
Take the Next Step Activity
Ethics and Technology Blog

8.5 Security Defenses Everybody Can Use

Whether protecting a large business or your personal laptop, there are certain common security defenses that help to prevent attacks and avoid data loss, including firewalls, software that detects and removes malware, and strong password protection.

Building Firewalls

As you learned in Chapter 6, a firewall is a part of your computer system that blocks unauthorized access to your computer or network, even as it allows authorized access. Firewalls can be created by using software, hardware, or a combination of software and hardware. On a Windows PC, the firewall is turned on by default and should always be left on. In rare cases, you may need to temporarily turn off the firewall to solve a software issue, such as installing a new program. Always turn the firewall back on immediately after the issue is resolved. On a Windows PC, you turn the firewall on and off in the Windows Defender Security Center.

Firewalls are like guards at the gate of the internet. Messages that come into or leave a computer or network go through the firewall, where they are inspected. Any message that doesn't meet preset criteria for security is blocked. You can set up trust levels that allow some types of communications and block others, or designate specific sources of communications that should be allowed access.

Keeping Viruses and Spyware under Control

All computer users should consider using **antivirus software** and **antispyware software** to protect their computers, data, and privacy. In fact, there are several free products that do a very good job, and antivirus/spyware functionality is often included within an operating system package. For example, Windows Defender, the Microsoft security suite, is built into Windows and has active, real-time protection turned on by default.

Antivirus products require that you update the **virus definitions** on a regular basis, ensuring that you have protection from new viruses as they are introduced. Antivirus products today generally install with settings turned on to automatically update the virus definitions on a regular basis, ensuring that you have constant protection from new malware threats. By default, these programs scan your system at regular intervals, usually during times when you are not using your PC. You can adjust the options for updates and scans and change the action that is performed when a virus is discovered. Once you have updated definitions, you run a scan and have several options: to quarantine viruses to keep your system safe from them, to delete viruses completely, and to report viruses to the antivirus manufacturer to help keep their definitions current. Most programs install with updates and scans set to be performed automatically at set intervals. Antispyware performs a similar function regarding spyware.

Most people use software that deals with both viruses and spyware in one package. Most are set to protect your computer in real time, meaning that they detect an incoming threat, alert you, and stop it before it is downloaded to your computer. Most programs also have anti-phishing detection built in. These provide alerts in email messages that flag dangerous links. Check the program you are using to make sure anti-phishing detection is on and active.

In addition to using antivirus and antispyware software, companies such as Microsoft release periodic updates that address issues in their software or new threats that have come on the scene since their operating system was released. Windows Update is set up to automatically download only those updates considered critical and alerts you to other updates that are

Windows Firewall blocks unwanted intruders from reaching your computer or network. This is an example of a software firewall.

Windows 10 installs with Windows Defender automatically turned on to secure your device from malware. Those using an older version of Windows (before Windows 8.1) can download and install the free Microsoft Security Essentials program that provides the same malware protection.

available. Apple provides updates and security enhancements in a similar fashion. When new software has been released to address a potential security threat, you are prompted in the update window that it is available to be installed.

Protect your smartphone or tablet with mobile antivirus and antispyware software. Computer security companies such as McAfee and Symantec offer mobile security apps. In addition to protecting you from malware, these apps often include a feature to lock and wipe your device remotely. Consider installing an antivirus program on your mobile device before you add any other apps to ensure that you're protected from the get-go.

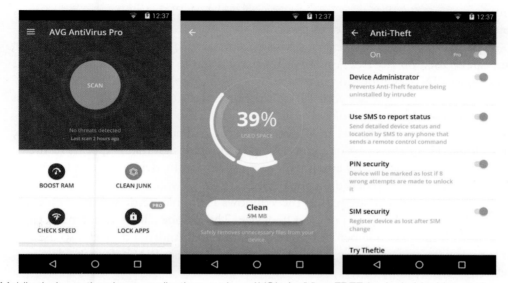

Mobile device anti-malware applications such as AVG's AntiVirus FREE for Android tablets and phones can be downloaded from the app store. Some companies (such as AVG) provide a free version and a paid version (sometimes called PRO). Paid editions provide additional features.

Using Passwords Effectively

Authentication is an excellent way to ensure that user access, often the weakest link and biggest risk to computer security, is managed effectively. However, a password-based authentication system is only as good as its users' passwords.

Passwords can be surprisingly weak and easily determined with automated password-breaking tools. For example, a dictionary attack can scan through every common word in the language in just minutes. The best system requires a user to create a **strong password** that combines uppercase and lowercase letters, numbers, and punctuation, and that does not include three or more repeating numbers or letters (333, for example) because these patterns are easy to discern.

In addition, a good password system requires that users change passwords frequently and provides password validation questions that are not easily discovered. For example, asking you to verify your place of birth or mother's maiden name is a weak system, because both facts are in the public record. Asking you your favorite vacation spot or piece of music is stronger because, although some close friends might know this information, the average criminal might have trouble figuring it out from publicly available records.

People should also create a system for themselves to keep track of passwords. Good passwords don't have to be hard to remember—just hard to guess. If you have a favorite phrase from a movie or song, type it in backwards and add a few numbers, rather than entering a random jumble of letters and numbers or your birthdate. The first is hard to remember, and the latter easy to guess. Also, it's okay to write down passwords, but don't keep them near your computer or mobile device where an intruder might find them.

Using PINs

Setting up a **PIN (personal identification number)** to access a device is becoming more commonplace. One difference between a PIN and a password is that the PIN is associated with the device; if you want to use the same PIN on another device, you must set it up on each laptop, smartphone, or tablet. A PIN is also shorter (as few as four digits) and therefore faster to enter.

A password is usually associated with an account, such as your Microsoft account, and can be used on any PC. A password is transmitted to a server for authentication, while most PIN authentication occurs using a cryptographic key on the local device. If someone steals or guesses your PIN, the person needs the physical device to access your data; if someone gets hold of your password, he can sign on to your account from anywhere on any device. In Windows 10, you can create a PIN that is up to six digits long. This PIN is part of a powerful encryption design that is specific to the device. After four failed PIN attempts, you must restart the device. After a few more failed tries, the PIN is blocked for the device.

PINs in Windows 10 are part of the Windows Hello biometric system. Apple uses similar passcode technology on any iOS device. The iPhone allows six attempts to sign on with the passcode before the phone is disabled. Android devices also use PINs to lock devices; you get to try 10 times to enter the correct PIN before the data is deleted and the phone reset.

A PIN isn't the most secure way to lock a mobile device. A four-digit PIN is easy to shoulder surf and someone sitting or standing nearby can easily figure out the digits you enter. Many people also use the same PIN on their smartphone, their ATM card, credit card, and home security alarm. Anyone intent on stealing your smartphone PIN may also gain access to your bank accounts or home.

Still, many people are fans of PINs because they're short and therefore easier to remember and enter. Security experts offer this advice: aim for six digits, don't choose something obvious such as a word spelled on the keypad, and don't use any of these PINs that account for nearly 20% of four-digit PINs: 1234, 1111, or 0000. Finally, use a different PIN on each mobile device and make sure it is not the same as what you use for your bank cards or security system.

> **Course Content**
> Take the Next Step Activity

Review and Assessment

Summing Up

8.1 The Role of Security and Privacy in Your Digital World

Computer security, also referred to as *information security*, involves protecting the boundaries of your home or business network and individual computing devices, including computers, tablets, and smartphones, from intruders (such as a malicious **hacker**). An important strategy is **data loss prevention (DLP)**, which involves minimizing the risk of loss or theft of data from within a network. Next-generation malware incorporates **advanced persistent threats (APTs)** where highly targeted, sophisticated attacks tailored to a specific organization or group of organizations are used to gain access to sensitive information.

The tools for protecting your personal computer and information from **identity (ID) theft** are similar to tools corporations use to protect their intellectual property: **authentication** (including **two-factor authentication, 2FA**), special software and hardware, and user procedures.

8.2 When Security Gets Personal

The increasing number of households connected to the internet has resulted in a growing number of cyberattacks, identity theft, and other types of fraud. In addition, loss of or damage to data resulting from harmful programs inadvertently downloaded to a computer can be frustrating and costly.

If you have set up a home network that connects to the internet, it is important to protect the network with a password.

Another important step in securing a home network is to use encryption. **Encryption** technologies protect stored data and data in transit. When you send an email, for example, you apply an **encryption key**. **Public key encryption** involves the use of a **public key** to encrypt data and a **private key** to decrypt it. Popular forms of encryption are **Wi-Fi Protected Access (WPA)**, **Wi-Fi Protected Access 2 (WPA2)**, **Wi-Fi Protected Access 3 (WPA3)**, and **Wired Equivalent Privacy (WEP)**.

It is also important to protect your computer against **malware**. Some common forms of malware are **viruses**, **worms**, **Trojan horses**, **macro viruses**, **logic bomb viruses**, **rootkits**, **botnets**, **spyware**, **adware**, **scareware**, and **ransomware**. Lately, more and more hackers are co-opting smaller IoT devices into their botnets. Malware is most often spread via the internet, so an important way to protect your computer is by knowing how to spot a trusted site and avoiding those that are known to spread malware. The ubiquitous presence of mobile devices in our daily lives has driven hackers to the world of mobile malware, so smartphones are at increasing risk. In addition, social media malware is becoming more common than traditional forms of malware infection.

Sites where you will perform financial transactions should always have **Transport Layer Security (TLS)** protection in place. TLS is a protocol—evolved from **Secure Socket Layer (SSL)**, an earlier protocol—that protects data such as credit card numbers as it is transmitted between a customer and an online vendor or payment company.

Other important skills include understanding how to manage **cookies** and not being caught by **phishing** scams. Hackers are now moving their phishing tactics to customers of cloud providers.

8.3 Mobile Security

Mobile computing carries its own security risks. There are several devices and procedures you can use to physically secure your mobile device including locks, fingerprint readers, and password protection. Companies may provide mobile computing policies for employees.

If you travel and access the internet using public hotspots (locations that offer Wi-Fi access), be very careful not to expose private information. Anything you send over a public network can be accessed by hackers and criminals. A type of attack that occurs in public areas with free Wi-Fi hotspots is known as a **man-in-the-middle (MITM) attack**, which allows a hacker to intercept data that is being sent from one device to another.

There are three major security issues when it comes to using mobile phones: first, protect the phone from theft; second, use caution if your phone features Bluetooth because nearby users can use this technology to track your communications or infect you with a virus; and third, avoid downloading malware which can result in the loss or theft of personal data. Also, as wearables become more commonplace, the amount of data being stored in the cloud and on smartphones raises the potential for privacy and security concerns.

Security concerns for IoT devices involve devices being taken over by cybercriminals intent on stealing data, creating harm to individuals, or adding to their botnet network. IoT devices may not receive frequent software updates to close loopholes as vulnerabilities are discovered.

8.4 Security at Work

Many corporations have sophisticated security strategies and tools in place to protect them against outside attacks and internal negligence by employees. The popularity of BYOD (bring your own device) policies in the workplace has brought new security risks to the attention of company owners and managers.

Using **symmetric encryption**, companies can protect stored data and data in transit on the internet or across a corporate network by designating which computers should be able to share encrypted data and then providing them with matching decryption keys.

Two tools used by businesses that have online transactions are digital signatures and digital certificates. A **digital signature** is a unique string of characters used to verify the authenticity of a file that travels to its destination via the internet. A **digital certificate** has been authenticated by a third party that validates the owner and issues the certificate.

An **Intrusion Prevention System (IPS)** is a robust anti-malware program that offers network administrators a set of tools for controlling access to the system. Many corporations also use audit tools to help spot random problems or attacks. One type of IPS called an **anomaly based intrusion detection system** can also detect suspicious content or unexpected traffic.

Stopping unauthorized access to the most sensitive company data may involve both physical security (for example, locking the door to the network server room) and a user authentication system. It is also important to provide employee training to help employees avoid phishing scams run by con artists, known as **social engineers**, who may try to talk employees into giving up corporate secrets or passwords.

Criminals constantly try to find ways around authentication systems that require user names and passwords. For example, **spoofing** is a technique used by malicious hackers to make it appear that they are someone else to convince a user to give up valuable information.

A **denial-of-service (DoS) attack** involves assaulting a corporate network with a constant stream of requests that causes the system or website to slow down or even crash.

Major damage of paper and electronic records can happen when an earthquake or hurricane hits. Companies typically have a **disaster recovery plan (DRP)** and backup systems using **cold servers**, **warm servers**, or **hot servers** to prevent the loss of data. Off-site storage is another good option for companies. Securing data in the cloud is also a concern for business owners and managers. To prevent data loss from power failures, a company uses **surge protectors** and **uninterruptible power supply (UPS)** devices.

Corporate IT departments may provide regular training for employees to keep them up to date on types of cyberattacks and the strategies to combat them.

8.5 Security Defenses Everybody Can Use

A firewall serves as a gatekeeper, blocking suspect access and allowing authorized access to your computer or network. Firewalls use software, hardware, or a combination of software and hardware to inspect incoming and outgoing communications. Any communication that doesn't meet preset criteria for security is blocked.

All computer users should use **antivirus software** and **antispyware software** to protect their computers, data, and privacy. Manufacturers of antivirus products provide regular **virus definition** updates to ensure that you have current information and therefore the best protection. In addition to using antivirus and antispyware software, companies like Microsoft release regular updates to their operating systems.

Some passwords can be easily determined with automated password-breaking tools. The best system requires a **strong password**, which combines uppercase and lowercase letters, numbers, and punctuation, and does not allow three or more repeating numbers or letters because these patterns are easy to discern. A good password system requires users to change passwords frequently and provides password validation questions that are not easily discovered.

Setting up a **PIN (personal identification number)** on a device is becoming more common. A difference between using a PIN and a password is that the PIN is associated with the device; if you want to use the same PIN on another device, you must set it up on each laptop, smartphone, or tablet. A password is associated with an account, like your Microsoft account, and can be used on any PC. PINs are shorter, and although they are quicker to enter, they can also be more easily guessed.

Terms to Know

8.1 The Role of Security and Privacy in Your Digital World

computer security, 209
data loss prevention (DLP), 209
hacker, 209
advanced persistent threat (APT)
    attack, 210

identity (ID) theft, 210
authentication, 211
two-factor authentication (2FA), 211

8.2 When Security Gets Personal

8.3 Mobile Security

8.4 Security at Work

8.5 Security Defenses Everybody Can Use

Projects

Check with your instructor for the preferred method to submit completed work.

Wireless Networks

Project 8A

Watch the video titled "Securing Your Wi-Fi" at https://ODW5.ParadigmEducation .com/SecuringYourWi-Fi. After watching the Google video, research Wi-Fi Protected Access 2 (WPA2), Wi-Fi Protected Access (WPA), and Wired Equivalent Privacy (WEP) wireless security standards. Prepare a brief summary comparing the three security protocols. Include the URLs of the source material you used.

Project 8B TEAM

Watch the video and read the article titled "Tips for Using Public Wi-Fi Networks" at https://ODW5.ParadigmEducation.com/PublicWi-FiTips and published by OnGuardOnline.gov. Develop a one-page flier that a travel agency could give to their customers that explains the risks involved with using public wireless hotspots to check their bank balance, send email, or update their social networks while traveling. Include tips for the traveler to securely communicate online with family members or friends while away from home. Make the flier appealing and easy to read. Be sure to write it in your own words. Submit the flier to your instructor.

Mobile Device Security

Project 8C

Read the article titled "10 Quick Tips to Mobile Security" at https://ODW5 .ParadigmEducation.com/10QuickTipstoMobileSecurity. Jump to page 7 in the PDF and then read the section titled "Top 10 Mobile Safety Tips." Prepare a one-page document in a checklist format that lists each of the ten tips. Next to each tip, provide a brief description in your own words. Then, using the checklist, interview a parent, friend, or relative who uses a smartphone for work and determine which tips the interviewee follows. After you have completed the checklist, ask the interviewee if his or her place of employment has a mobile security policy. If so, modify the checklist to include any tips or practices that are not on the original form. Prepare a summary of the interview that details which practices were and were not followed. Submit the completed checklist and summary to your instructor.

Project 8D TEAM

A local company has hired your team to consult on a new mobile device security policy to protect its IT assets. The company employs five consultants who travel around the country conducting training presentations on individual taxation reporting requirements. Each consultant travels with a company smartphone, laptop, and in some cases a tablet and is typically away from the office for three days. To prepare you for this consultation, read the slides in the presentation titled "The Billion Dollar Lost-Laptop Study" at https://ODW5.ParadigmEducation.com/LostLaptopStudy as a starting point and then research examples of mobile security policies. Discuss with your team the main elements that should be covered in a mobile device security policy to safeguard the company from the disruption in service and large costs associated with lost mobile devices. Prepare a presentation for the company's executive board that outlines the main points that your team recommends be included in the mobile device security policy for smartphones, laptops, and tablets. Include a slide at the end of the presentation with links to the sources used for sample policies.

Project 8E

The popularity of smart devices connected in the Internet of Things is raising security and privacy alarms for many computer experts. At a recent consumer electronics conference in Berlin, Marco Preuss, a director at security company Kaspersky Labs, said that third party access to data transmitted by smart appliances should be cause for concern. One example he gave was a health insurance provider finding out what you have stored inside your fridge and upping your policy cost because of your poor eating habits. Research security and privacy settings for smart appliances and learn what a consumer can do to secure them from hackers and keep information private. Prepare a bullet-point summary of what you learned. Include the URLs of the articles you read.

Thwarting Social Engineers

Project 8F

You have been asked by your company's human resources department to prepare a training proposal for employees on the security threats posed by social engineers. Read the article titled "Francophoned: A Sophisticated Social Engineering Attack" at https://ODW5.ParadigmEducation.com/Francophoned. This article, published by Symantec, provides details about the use of emails and telephone calls by cybercriminals that led to money being stolen from accounts. Find two additional articles that describe the manipulative practices of social engineers. Next, consider the information that needs to be covered in your employee training program in order to safeguard your company's confidential information. What topics should be included in the training? What would be the best training method? How can you test employee understanding of the training and compliance with proper security protocols and procedures? Prepare a training proposal that includes a rationale for the training, an outline of the content, a description of the methodology you would use, and suggestions for testing employee knowledge and compliance. At the end of the proposal, include references to the two additional articles you read for this project.

Project 8G TEAM

Your IT manager has decided she wants to perform a test to evaluate the company's recent social engineering training. The manager has targeted two positions that she believes are the most vulnerable to social engineers: the help desk staff and the front desk receptionists. Develop a script for a telephone conversation in which a social engineer calls the help desk and attempts to get a staff member to reveal the user name and reset the password of an employee. Develop another script for a face-to-face conversation between a receptionist and a social engineer who tries to gain entry to the locked server room by impersonating an outside contractor.

Project 8H TEAM

Your team has volunteered to help senior citizens at a local community center. Recognizing that this age group is easily victimized by social engineers, your team would like to inform them about the manipulative practices of social engineers and the ways in which they can protect themselves. Develop a presentation that demonstrates the tactics that a social engineer uses to trick senior citizens into sending money to cyber-criminals or into revealing private information. Include tips for seniors on how to avoid being scammed.

The Best Defense Is a Good Offense

Project 8I

Research two to three free mobile security apps that guard against malware and hackers for your smartphone or tablet. If you don't own a smartphone or tablet, research apps for Android, the most widely used mobile operating system. Read reviews of the apps from a credible source. Create a document that summarizes in bullet points and in your own words what you learned about each app and the reviewers' feedback. Conclude by indicating the app you would download and install to protect your mobile device and give your reasons for your choice. Include the URLs of articles that you read for this project at the end of the document.

Project 8J TEAM

Your team has been hired to consult about security with an entrepreneur who is opening a new catering business. The owner has told your team she has a business manager who will provide price quotes and meet with customers to plan menus, an executive chef who will be responsible for all food and beverage operations, and an administrative assistant who will liaise with all internal and external stakeholders for catering events. These key people will all have company-provided smartphones and tablets because they are expected to communicate from event locations. They will use a Wi-Fi network in the office to connect office computers with their mobile devices and the office printer. As a team, identify and discuss the types of security risks that the owner will have to guard against. Prioritize each risk using the scale High Risk, Medium Risk, and Low Risk. Create a presentation for the owner that describes each risk. Include references at the end of the presentation for any articles that you use to help you with this project.

Project 8K

Go to a website where you've recently shopped online (or any online retailer) and try to locate their privacy policy. Read the privacy policy and then prepare a summary with answers to the following questions.

- What personal information about customers is collected by each site?
- How is customer personal information kept secure?
- What choices do customers have with respect to their information?

Submit the URL for the privacy policy you read with your answers.

Wiki Project—Computer Security

Project 8L `TEAM`

You or your team will be assigned to research and write a response to one or more of the following FAQs associated with internet security and safety:

- How can I protect my mobile device from malware?
- How can I tell if an email message is a phishing attempt?
- How can I tell if a text message I've received is a phishing attempt or a vehicle for a hacker to send me malware?
- How do I protect my personal information and/or privacy on a social networking site such as Facebook?
- How can I tell if a checkout page is secure?
- How do I turn on the lock/passcode feature on my tablet?
- How do I hide my mobile device when Bluetooth is on?
- How can I control a website's placement of a cookie on my computer?
- How can I tell if my computer has a virus?
- How do I secure my wireless network at home?
- How can I make sure that all of my personal information is cleaned from the computer, smartphone, or tablet that I am donating to charity?
- What can I safely do online at a cyber café or other public Wi-Fi hotspot?

Post to the course wiki site your response for the question(s) assigned to you. In the post, include your response and any examples that would aid comprehension. Be sure to include references to the sources of your information.

Project 8M `TEAM`

Your team is assigned to edit and verify the content posted on the wiki site from Project 8L. If you add or edit any content, make sure you include a notation within the page that includes your name or the team members' names and the date you edited the content—for example, "Edited by [your name or team members' names] on [date]." Keep a list of references that you used to verify your content changes. When you are finished with your verification, include a notation at the end of the entry—for example, "Verified by [your name or team members' names] on [date]."

Class Conversations

Topic 8A   What is the role of employees in keeping data secure?

A cybersecurity firm hired by a company to analyze security risks discovers that the weakest link is the company's own employees. The consultant says that the biggest leaks of sensitive company data occurred while employees were outside having a cigarette or down at the local pub in the evening, where they openly talked about private, sensitive information. Others could easily overhear this information and use it against the company in a targeted attack. Discuss strategies an employer can adopt to convince employees that sensitive company information is not to be discussed outside the office building, on social networking sites, or in other personal communications. Would it be hard to convince employees to embrace these strategies? Why or why not?

Topic 8B Do you care if advertisers are in your social network space?

Discuss the pros and cons of social networks targeting advertising toward you based on your personal data, buying history, and web browsing practices outside the social network. Do you like that the advertising you see is customized to your life, or does it bother you that businesses are exchanging data about your online presence without your knowledge? Why or why not?

In 2018 it was revealed that the personal data collected by Facebook was readily available and misused by third-party company Cambridge Analytica. This data may have been used to influence the 2016 presidential election. Did this revelation cause you to re-examine your privacy settings? Why or why not?

Topic 8C Stop that data from walking out the door!

Read the article titled "Employee Termination from an IT Perspective" at https://ODW5.ParadigmEducation.com/EmployeeTermination. Do you believe employees who leave a company are entitled to take some data away with them? If yes, what data would be acceptable to take and what data would not be acceptable to take? Is a non-disclosure agreement really effective in protecting company data? Why or why not?

Glossary

3-D printing The process of using a specialized printer to create 3-D objects from digital models. Ch 3

3G The third generation of wireless telecommunication standards and technology. Ch 4

4G The fourth generation of wireless telecommunication standards and technology. Ch 4

5G The fifth generation of wireless telecommunication standards and technology. Ch 4

4K A technology that provides more vivid images for TVs, cameras, and monitors with almost twice the resolution of FullHD displays. Ch 3

802.11 standard A communications standard used in Wi-Fi networks that tells wireless devices how to connect with each other using a series of access points and radio frequencies to transmit data. Ch 6

A

access speed The speed with which data in RAM can be accessed and processed by a computer. Ch 3

advanced persistent threat (APT) A highly targeted, sophisticated attack tailored to a specific organization; usually done to gain access to sensitive information. Ch 8

adware Software that is supported by advertising and is capable of downloading and installing spyware. Ch 8

alpha version The first stage of testing a software product. Ch 5

American National Standards Institute (ANSI) One of the organizations that establishes network communications standards. Ch 6

analog computer A computer that uses mechanical operations to perform calculations, as with an older style car speedometer or slide rule. Ch 1

analog signal An electronic signal formed by continuous sound waves that fluctuate from high to low. Ch 6

Android An open source operating system initially developed by Google that runs mobile devices such as smartphones, tablets, and netbooks. Ch 4

animation software Software used to create a series of moving images. Ch 5

anomaly based intrusion detection system One type of intrusion prevention system that detects suspicious content or unexpected traffic using rules (heuristics) about normal system operation rather than patterns or signatures (as with antivirus software). Ch 8

antispyware software Software used to prevent the downloading of spyware to a computer or network, or to detect and delete spyware on the system. Ch 8

antivirus software Software used to prevent the downloading of viruses to a computer or network, or to detect and delete viruses on the system. Ch 8

application software Includes the software products you use to get tasks done, from producing reports for work to creating art or playing games. Ch 5

artblog A blog where people post creative works such as music, photos, or paintings. Ch 7

ARM Previously called *Advanced RISC (reduced instruction set computing) Machine*. A technology that contains a simplified architecture for a computer process to increase battery life. Ch 1

artificial intelligence (AI) The ability of computing devices to exhibit intelligence. Ch 1

ASCII (American Standard Code for Information Interchange) A computer code that numerically represents letters in the English alphabet to enable the transfer of text between computing devices. Ch 1

assistive technology A device that enables physically challenged computer users to control their computers through actions such as puffing or using a stick. Ch 3

audio conferencing Transmission technology that allows you to make voice calls over the internet. Ch 2

audio software Software that allows you to record and edit audio files. Ch 5

augmented reality Adding overlays of technology, such as audio or video, to improve our real-world experiences. Ch 3

authentication The use of passwords or other identifiers, such as fingerprints, to make sure that the people accessing information are who they claim to be and have permission to get such access. Ch 8

B

backbone A pathway for transfers of information in a network. Ch 6

bandwidth The number of bits (pieces of data) per second that can be transmitted over a communications medium. Ch 6

bar code reader An input device that optically scans a series of lines, or a bar code, using a light beam. Ch 3

beta version The second stage of testing a software product. Ch 5

binary digits A value in the binary system, either 1 or 0. Ch 1

binary system A system consisting of two possible values, 0 and 1, called *binary digits*, or *bits*. Ch 1

biometrics Technology that uses devices such as fingerprint readers or retinal scanners to identify a person by a unique physical characteristic. Ch 8

BIOS (basic input/output system) Code that checks and starts up computer devices such as memory, monitor, keyboard, and disc drives, and directs the hard drive to boot up and load the operating system (OS) to memory. Ch 3

bit The smallest unit a computer can understand and act on. An abbreviation for *binary digit*. Ch 1

blog An online journal; short for *web + log*. Ch 7

bluejacking When a hacker uses your Bluetooth connection to send information to another Bluetooth device. In a bluejacking attack, the connectivity is being compromised but your personal data may not be accessed. Ch 8

bluesnarfing When a cybercriminal gains access to your smartphone via your Bluetooth connectivity with the intent to steal all your personal data and files on your phone. Ch 8

Bluetooth A network protocol that offers short-range connectivity (3 to 300 feet, depending on a device's power class) via radio waves between devices such as cell phones and cars. Bluetooth-enabled devices can communicate directly with each other. Ch 6

Bluetooth 5 Latest Bluetooth standard; became available in 2017. Ch 6

Bluetooth headset A wireless headset that can receive data (including sound) from Bluetooth-enabled devices. Ch 3

Bluetooth LE (BLE) Bluetooth low energy (also referred to as *Bluetooth Smart*) is a new Bluetooth standard that uses much less power to communicate within the same range as what is now called "Classic" Bluetooth. Ch 6

Bluetooth Smart Devices that use the Bluetooth low energy standard (also referred to as *Bluetooth LE*) are marketed under the name Bluetooth Smart. These devices use much less power to communicate within the same range as what is now called "Classic" Bluetooth. Ch 6

Blu-ray disc A type of storage device, mostly used for high-definition movies and games. Ch 3

bookmark A way to save a reference to a website or web page so that it can be easily visited again. Ch 7

booting The process of starting your computer. Ch 4

botnet A group of computers that have been compromised (*zombies* or *bots*) so they can forward communications to a controlling computer. Ch 8

bridge A device that helps separate, similar networks to communicate with each other. Ch 6

broadband Any communications medium that is capable of carrying a large amount of data at a fast speed. Ch 6

browser An application used to view pages and content on the web, and to navigate among pages. Ch 2

bug A computer industry term for a flaw or failure in software that causes it not to work as intended. Ch 5

bus A subsystem that moves instructions and data around the components on a computer. Ch 4

bus topology An arrangement where all the computers and other devices on a network, such as printers, are connected by a single cable. Ch 6

business-to-business (B2B) e-commerce A product, service, or payment between two online businesses. Ch 2

business-to-consumer (B2C) e-commerce A product, service, or payment between an online business and individual consumers. Ch 2

BYOD (bring your own device) The practice of employees bringing their own devices to the workplace and connecting the devices to company networks. Ch 4

byte A collection of 8 bits. Ch 1

C

cable modem A piece of hardware that enables you to send and receive digital data using a high-speed cable network based on the cable television infrastructure found in many homes. Ch 6

cache memory A memory area, located on or near the microprocessor chip, where the most frequently used data is stored. Ch 1

calendar software Software designed to schedule appointments or events and to set up reminders. Ch 5

CD A storage device from which a computer can read data, write data, or both. Ch 3

cellular network A transmission system that sends signals through a cell tower, used to transmit both voice and data in every direction. Each cell tower has its own range, or cell, of coverage. Ch 6

cellular transmission A signal sent using a cell tower. Ch 6

central processing unit (CPU) The part of your computer system that interprets instructions and processes data. Also referred to as the *processor* or *core*. Ch 1

charging mat A device that allows you to charge multiple devices at once without connecting them to the mat with a cable. Also called a *charging pad*. Ch 3

chip A thin wafer of semiconducting material, such as silicon, which contains an integrated circuit that performs various functions, such as performing mathematical calculations, storing data, or controlling computer devices. Ch 3

Chrome OS Developed by Google and based on Linux, this operating system is designed to have specific functionality delivered via web-based apps. Ch 4

client A computer or other device capable of sending data to and receiving data from a server on a network. Ch 6

client/server network A network architecture in which a computer (called the *server*) stores programs and files that any connected device (called a *client*) can access. Ch 6

clock speed The speed at which a processor can execute computer instructions, measured in gigahertz. Ch 3

cloud computing A model of software delivery in which software is hosted on an online provider's website and you access it over the internet using your browser; you don't have to have the source application software installed on your computer in order to use the software. Also called *utility computing*. Ch 5

cloud storage Services that allow users to store documents and other files online. Ch 1

coaxial cable The cable used to transmit cable television signals over an insulated wire at a fast speed (millions of bits per second). Ch 6

cold boot Starting a computer from a no-power state. Ch 4

cold server A spare server used to take over server functions. Ch 8

command-line interface Refers to operating systems (such as DOS) where the user had to type in text commands. Ch 4

comma-separated-values (CSV) format A format that separates each piece of data in a file with a comma. Ch 5

communications system In the context of a computer network, a system that includes sending and receiving hardware, transmission and relay systems, common sets of standards so all the equipment can "talk" to each other, and communications software. Ch 6

computer An electronic, programmable device that can assemble, process, and store data. Ch 1

computer cluster A group of computers joined together to provide higher computing power. Ch 1

computer engineering (CE) The study of computer hardware and software systems and programming how devices interface with each other. Ch 1

computer memory Temporary storage areas on your computer, including random access memory (RAM) and cache memory. Ch 1

computer network Two or more computing devices connected by a communications medium, such as a wireless signal or a cable, that is managed via the operating system to share resources. Ch 6

computer science (CS) The study of how to design software, solve problems such as computer security threats, or come up with better ways of handling data storage. Ch 1

computer security Activities that protect the boundaries of individual computing devices and home and business networks from intruders. Ch 8

computing power Tasks accomplished by a computing system compared to the resources used. Ch 1

consumer-to-consumer (C2C) e-commerce A product, service, or payment between two online consumers. Ch 2

contact management software A type of software used to store, organize, and retrieve contact information. Ch 5

converged device A type of device that combines several technologies, such as the ability to calculate, store data, and connect to the internet. Examples of this type of device include smartphones, GPS navigation systems, digital cameras, and smart appliances. Ch 1

cookie A small file stored on your computer by a web server to track information about you and your activities. Ch 8

copyright Legal ownership of a work or symbol. Ch 2

Cortana The personal assistant/search feature introduced in Windows 10. Ch 4

crawler An intelligent agent that follows a trail of hyperlinks to locate online data. Ch 2

customer relationship management (CRM) software Generally, a suite of software or online services used to store and organize client and sales prospect information, and automates and synchronizes other customer-facing business functions, such as marketing, customer service, and technical support. Ch 5

cyberforensics A field of study or a career that involves extracting information from computer storage that can be used to provide evidence in criminal investigations. This might involve decrypting encrypted data or finding residual data on a hard drive that somebody has tried to erase. Ch 8

cybersecurity Methods used to protect the computing environment of an individual user, an organization, or a government. Ch 1

D

data Raw facts; what you put into a computer. Ch 1

data integration Combining data from several online sources into one search result. In the context of online search, cross referencing various sources of data and including those references in the results. Ch 2

data loss prevention (DLP) Activities that involve minimizing the risk of loss or theft of data from within a network. Ch 8

database administration A career field in which workers are responsible for making sure that information stored in a database is available to and usable by those who need access to the information and that the information is secure from unauthorized access. Ch 1

database software A type of software used to query, organize, sort, and create reports on sets of data such as customer lists. Also called *database management system (DBMS)*. Ch 5

deep web Databases and other content on the web that aren't catalogued by most search engines. A typical search engine won't return links to these databases or documents when you enter a search keyword. Also called the *invisible web*. Ch 2

denial-of-service (DoS) attack An attack against a corporate system that slows performance or brings a website down. Ch 8

desktop The background image shown on the screen upon which graphical elements such as icons, buttons, windows, links, and dialog boxes are displayed. Ch 4

desktop computer A non-portable computer whose central processing unit (CPU) is housed in a tower configuration or, in some cases, within the monitor, as with the Apple iMac. Ch 1

desktop publishing (DTP) software Software used to lay out pages for books, magazines, and other print materials such as product packaging or brochures. Ch 5

dial-up modem A piece of hardware that works with telephone transmissions and changes or manipulates an analog signal so that it can be understood by a computer or fax machine (which only understand digital signals). Ch 6

digital certificate An electronic document used to encrypt data sent over a network or the internet. Ch 8

digital computer A computer that represents data using binary code. Ch 1

digital pen A device used to write or draw on a touchscreen. Ch 3

digital signal A discrete electronic signal that is either high or low. In computer terms, high represents the digital bit 1; low represents the digital bit 0. Ch 6

digital signature A mathematical way to demonstrate the authenticity of a digital certificate. Ch 8

disaster recovery plan (DRP) A formal set of policies and procedures that guides the preparation for a possible disaster and subsequent recovery of computer resources and information thereafter. Ch 8

Disk Cleanup A utility included with Windows that gets rid of unused files on your hard disk. Ch 4

disappearing media Social media posts that can be viewed for only a short amount of time before they disappear and are deleted from the site's servers; a trend driven by social apps such as Snapchat. Ch 7

distributed application architecture A network architecture that distributes tasks between client computers and server computers. Ch 6

document camera An output device that is often used in educational settings to display text from a book, slides, a 3-D object, or any other printed material. Ch 3

domain name An identifier for a group of servers (the domain) hosting a particular website. Ch 2

DOS An early operating system for personal computers which used a command-line interface. Ch 4

download To transmit data, such as a digitized text file, sound, or picture from a remote site to one's own computer via a network, such as the internet. If you are receiving content from the internet, you are downloading. Ch 2

DRAM (dynamic random access memory) The type of memory most commonly found in computers; works quickly and is compact and affordable. Requires electricity and is fragile, meaning that the data held in RAM must constantly be refreshed. Ch 1

drive A device that stores data on media. Can be integrated into the computer, be external, or removable. Ch 3

driver Software that allows an operating system to interface with specific hardware, such as a printer or keyboard. Ch 4

DSL modem A piece of hardware that allows you to connect to your existing telephone system but separates voice from data traffic so you don't lose the use of your telephone while your computer is transmitting or receiving data. DSL stands for *digital subscriber line*. Ch 6

DVD A storage device from which you can read data, write data, or both. This type of media can store larger quantities of data than a CD. Ch 3

E

e-commerce The transaction of business over the internet. Ch 2

edge computing Use of a locally based (close to where the data is generated) data center to process critical data and then send batches of the processed data to a cloud provider for storage. Ch 6

edutainment Software programs and/or media that contain both entertainment and educational value. Ch 5

email (electronic mail) A message that is shared over the internet. Ch 2

email address A unique identifier for an email sender or recipient, comprised of user name, domain name of the email service, and domain suffix. Ch 2

email client A program stored on your local computer that is used to manage multiple email accounts and contacts when you connect to your email service. Ch 2

embedded object An object that has been inserted into a software file using object linking and embedding (OLE) technology, where the object is not connected to the object in the source file and therefore will not be altered when the source file is changed. Ch 5

encryption The process of using a key to convert readable information into unreadable information to prevent unauthorized access or usage. Ch 8

entertainment software A category of software that includes computer games you play on a computer or game console. Ch 5

Ethernet A standard that specifies that there is no central device controlling the timing of data transmission. With this standard, each device tries to send data when it senses that the network is available. Ch 6

extension A program that runs in a browser and extends the functionality of the browser. Ch 2

expansion card A device inserted on the computer motherboard that adds capabilities such as sound, graphics handling, or network communications. Ch 3

export To send data to another document. Ch 5

external hard drive A disk drive that connects to your computer via a cable connected to a port; allows you to store data and retrieve it from another computer. Ch 3

extranet An extension of an intranet that allows interaction with those outside the company, such as suppliers and customers. Ch 6

F

failover The process of redirecting users to a hot server. Ch 8

fax machine A device used to transmit a facsimile (copy) of a document to another location using a phone line. Ch 3

fiber-optic cable A transmission medium that uses a protected string of glass that transmits beams of light. Fiber-optic transmission is very fast, sending billions of bits per second. Ch 6

field A column of data in a table, which contains similar information. Ch 5

file A computer's basic storage unit; may contain a report, spreadsheet, or picture, for example. Ch 1

file allocation table (FAT) A table maintained by the OS to keep track of the physical location of the hard disk's contents. Ch 4

file extension The part of a file name that identifies the program that is to launch when the file is double-clicked. It is commonly a set of 3 or 4 characters following a period at the end of the file name. For example, for the file *index.htm*, "index" identifies the file and "htm" is the extension, indicating in this case that a browser such as Internet Explorer would launch to view the HTML (web page) file. Ch 5

File Transfer Protocol (FTP) A protocol that permits the upload and download of files. Ch 2

FireWire Port Based on the same serial bus architecture as a USB port, this type of port provides a high-speed serial interface and is generally used for devices that require high performance, for example, for digital cameras, camcorders, or external hard disk drives. Ch 3

firewall Software and hardware systems that stop those outside a network from sending information into the network or taking information out of it. Ch 6

Firewall as a Service (FWaaS) An approach to network security that connects an entire organization's network traffic from all locations and all users into a cloud infrastructure, where the cloud's firewall enforces the security policies. Ch 6

firmware Code built into electronic devices that controls those devices and may include instructions to start the system. Ch 4

flash drive A small, convenient device that lets you store data and take it with you. Also known as a *USB stick* or *thumb drive*. Ch 3

flash memory A type of computer memory used to record and erase stored data and transfer data to and from your computer; used in mobile phones and digital cameras because it is much less expensive than other types of memory. Ch 3

flop A measurement of computing power representing one floating-point operation per second. Ch 1

freeware Software that is made available to use free of charge. Ch 5

frequency The speed at which a signal can change from high to low; a signal sent at a faster frequency provides faster transmission. Ch 6

friends list The people you have allowed to access your profile on a social network. Ch 7

G

gaming device A piece of equipment, such as an Xbox, that allows a user to play a computer game using software or an online connection. Ch 3

gateway A device that helps separate, dissimilar networks to communicate with each other. Ch 6

generic top-level domain (gTLD) Common top-level domains with three or more characters, in contrast to two-character country code TLDs. Common gTLDs include .com, .org, and .net. Ch 2

gigabit per second (Gbps) A transmission at a rate of 1 billion bits per second. Ch 6

gigabytes (GB) The average computer's hard drive capacity for data storage is measured in gigabytes; one gigabyte is approximately 1 billion bytes. Ch 3

gigahertz (GHz) A measurement of processor speed; one gigahertz is approximately 1 billion cycles per second. Ch 3

GNU General Public License (GNU GPL) A policy that specifies polices about creating open source software, including that source code has to be made available to all users and developers. Ch 5

graphical user interface (GUI) The visual appearance of an operating system that uses graphical icons, buttons, and windows to display system settings or open documents. Ch 4

graphics software Software that allows you to create, edit, or manipulate images. Ch 5

H

hacker A person who has knowledge of computer technology and security settings and uses it for benign or malicious purposes. Ch 8

handwriting recognition software A type of software that enables a computer to recognize handwritten notes and convert them into text. Ch 5

hard disk The disk that is built into your computer and is the primary method of data storage. The disk rotates under a read/write head that reads and writes data. Ch 3

hardening Using a combination of hardware, software, and computer-user policies to make a network more resistant to external attack. Ch 6

hashtag The pound symbol (#); placed in front of topic names on social media sites as a way to help organize content and make it searchable. Ch 7

high dynamic range (HDR) A technology that provides more vivid photos with almost twice the resolution of standard photos. Ch 3

home page The main page of a website or the page a browser first lands on when you specify a universal resource locator (URL). Ch 2

honeypot As part of a corporate security strategy, a computer set up to be easily hacked into to help identify weaknesses in the system. Ch 8

hot server A spare server that receives frequent updates and is available to take over if the server it mirrors fails. Ch 8

hotspot A location where Wi-Fi access is available. Ch 6

hybrid cloud An information technology system in which some computing resources and data are managed internally in conjunction with other computing resources and data that are managed externally by cloud service providers. Ch 6

hyperlink Describes a destination within a web document and can be added to text or a graphical object such as a company logo. Used for navigation. Generally, clicking on a hyperlink sends the user to the specified web document. Ch 2

hypertext Text that represents a hyperlink. Used for navigation. Generally, clicking on hypertext sends the user to the specified web document. Ch 2

I

identity The profile you create when you join a social networking service. Ch 7

identity (ID) theft Taking someone else's personal information to falsely obtain funds, cred, or other advantages. Ch 8

import To bring content into a document. Ch 5

information Raw facts that are processed, organized, structured, or presented in a meaningful way; what you get out of a computer. Ch 1

information processing cycle A cycle of handling raw data and information that has four parts: input of data, processing of data, output of information, and storage of data and information. Ch 1

information systems (IS) A computer profession that bridges the needs of an organization and the way their information is handled to solve business problems. An IS professional considers who needs what data to get work done and how it can be delivered most efficiently. Ch 1

information technology (IT) The study, design, development, or management of computer systems, software applications, and computer hardware. Ch 1

Infrared Data Association (IrDA) port A port that allows you to transfer data from one device to another using infrared light waves. Ch 3

infrared (IR) technology A technology that enables transfer of data over short distances using light waves in the infrared spectrum. Ch 3

input Data that is entered into a computer or other device, or the act of reading in such data. Ch 1

input device A device that allows a user to put data into a computing device. Translates into electronic (digital) form. Ch 3

Institute of Electrical and Electronics Engineers (IEEE) One of the organizations that establishes network communications standards. Ch 6

instant message (IM) A message transferred over a network and requiring that the sender and receiver have the same messaging software. Ch 2

instruction register A holding area on your computer where instructions are placed after the fetch portion of the machine cycle is completed. Ch 1

intellectual property Creations of the mind; inventions, literary and artistic works; and symbols, names, images, and designs. Ch 2

interactive whiteboard (IWB) A display device that receives input from the computer keyboard, a stylus, a finger, a tablet, or other device. Ch 3

internet The physical infrastructure that provides us with the ability to share resources and communicate with others across a network of computers. The internet is made up of hardware such as servers, routers, switches, transmission lines, and towers. Ch 2

Internet Message Access Protocol (IMAP) A communications protocol that receives email from a mail server and delivers it to the proper mailbox. Has replaced POP on some email servers. Messages will not be deleted from the server until requested by the recipient. Ch 2

Internet of Things (IoT) The network of physical devices embedded with technology that enable these objects to connect and exchange data over a network. Ch 2

internet peer-to-peer (P2P) network A modification of the peer-to-peer network architecture, used on the internet to share files. Ch 6

Internet Protocol (IP) address A series of numbers that uniquely identifies a location on the internet. An IP address consists of four or eight groups of numbers separated by a period, for example: 225.73.110.102. Ch 2

internet service provider (ISP) A company that lets you use its technology to connect to the internet for a fee, typically charged monthly. Ch 2

intranet A private network within a company's corporate "walls." Ch 6

Intrusion Prevention System (IPS) A robust form of anti-malware program that offers network administrators a set of tools for controlling access to the system and stopping attacks in progress. Ch 8

invisible web Databases and other content on the web that aren't catalogued by most search engines. A typical search engine won't return links to these databases or documents when you enter a search keyword. Also called the *deep web*. Ch 2

iOS The Apple operating system for the iPhone, iPad, iPod Touch, and Apple TV devices. Ch 4

K

keyboard An input device that consists of keys a user presses to input data. Ch 3

keystroke logging software A kind of malware that is used to track the keystrokes typed by a computer user. Ch 3

keyword Word or phrase that you include in search text to look for content using a search engine. Ch 2

kilobit per second (Kbps) A transmission at a rate of 1 thousand bits per second. Ch 6

L

laptop A portable computer with a built-in monitor, keyboard, and pointing device, along with the central processing unit (CPU) and a battery. Also known as a *notebook computer*. Ch 1

LCD projector A device that projects light through panels made of silicone colored red, green, and blue. The light passing through these panels displays an image on a surface such as a screen or wall. Ch 3

least possible privileges A principle applied by network operating systems that means that each user is only given access to what he or she needs in order to get his or her work done. Ch 4

light-emitting diode (LED) display A type of monitor that uses light-emitting diodes, saving power and delivering a high-quality image. Ch 3

linked object An object that has been inserted into a software file using object linking and embedding (OLE) technology, where the object is connected to the object in the source file and therefore is altered when the source file is changed. Ch 5

Linux First developed by Linus Torvalds in 1991, Linux is an open source operating system, meaning that the source code for it is freely available for use and modification. Ch 4

live broadcast Live, or *real-time*, delivery of media over the internet. Also called *live media streaming*. Ch 7

live media streaming Live, or *real-time*, delivery of media over the internet. Also called *live broadcast*. Ch 7

local area network (LAN) A type of network where connected devices are located within the same room or building, or in a few nearby buildings. Ch 6

logic bomb virus A piece of code that is placed in a software system to set off a series of potentially damaging events if certain conditions are met. Ch 8

Long Term Evolution (LTE) A set of 4G wireless standards that involves changes to the wireless infrastructure to increase speed and bandwidth by installing transmitters that operate on different frequency bands to avoid interference. Ch 6

M

machine cycle A cycle a CPU goes through when handling an instruction; a process in which four basic operations are performed: (1) fetching an instruction, (2) decoding the instruction, (3) executing the instruction, and (4) storing the results. Ch 1

macOS The operating system produced by Apple Inc. Ch 4

macro virus A form of virus that infects the data files of commonly used applications such as word processors and spreadsheets. Ch 8

malware Collectively, damaging computer programs such as viruses and spyware, which can do anything from displaying pop-up window advertisements to destroying your data or tracking your online activities. Ch 8

man-in-the-middle (MITM) attack A type of attack that occurs in public areas with free Wi-Fi hotspots that allow a hacker to intercept data being sent from one device to another. Ch 8

media sharing Sharing video, photos, music, or presentations with individuals or groups using the internet. Ch 7

media sharing site A service that allows you to share media, such as music and video, on the web. Ch 7

megabit per second (Mbps) A transmission at a rate of 1 million bits per second. Ch 6

megahertz (MHz) A measurement of RAM access speed; 1 megahertz is approximately 1 million cycles per second. Ch 3

memory capacity The amount of memory (RAM) on your computer; used to run programs and store data. Ch 3

mesh topology A network topology in which all devices in the network are interconnected. Ch 6

metadata Data about other data; describes that data and how to process it. Ch 7

metasearch engine A type of search engine that can search for keywords using several search engines at the same time. Ch 2

metropolitan area network (MAN) A type of network that connects networks within a city or other populous area to a larger high-speed network; typically made up of several LANs that are managed by a network provider. Ch 6

microblogging A form of blogging where brief comments rather than personal blogs are the main form of interaction, as on Twitter. Also called *social journaling*. Ch 7

microphone An input device for sound. Ch 3

microprocessor A computer chip that can accept programming instructions that tell a computer what to do with the data it receives. Ch 1

Microsoft Windows The operating system produced by Microsoft Corporation, first released in 1985. Ch 4

microwave A high-frequency radio signal that is sent from one microwave tower to another. Because the signal cannot bend around obstacles, the towers have to be positioned in line of sight of each other. Ch 6

MID (mobile internet device) A category of devices that fall between netbooks and phones, putting the internet in a pocket-sized form. Ch 3

MIDI A protocol that allows computers and devices, such as musical synthesizers and sound cards, to control each other. Ch 3

mobile app development A career that involves designing and creating apps for mobile devices such as smartphones and tablets. Ch 1

mobile broadband stick A USB device that acts as a modem to give your computer access to the internet; can easily be moved between computers. Ch 6

mobile forensics The field of study or career that involves finding data saved or sent via a mobile device to use as evidence in criminal prosecutions. Ch 8

mobile instant messaging (MIM) A message that is transferred by a mobile device over a Wi-Fi network and requires that the sender and receiver have the same messaging software. Ch 2

mobile operating system Operating system used on mobile phones and tablets. Often called *mobile OS* or *mobile platform*. Ch 4

modem A piece of hardware that sends and receives data from a transmission source such as your telephone line or cable television connection. The word modem comes from a combination of the words *modulate* and *demodulate*. Ch 6

monitor A visual output device that displays data and information as well as provides the ability to view the computer's interface. Ch 3

MOOC (massive open online course) A free online course typically with open registration and publicly shared curriculum; objective driven instruction with required assessments and some student-student and student-instructor interaction. Ch 5

Moore's Law A theory proposed by Gordon Moore, one of the founders of Intel, which states that over time the number of transistors that can be placed on a chip will increase exponentially, with a corresponding increase in processing speed and memory capacity. Ch 3

motherboard The primary circuit board on your computer that holds the central processing unit (CPU), BIOS, memory, and expansion cards. Ch 3

mouse An input device, also referred to as a *pointing device*, that is able to detect motion in relation to the surface you rest it on and provides an onscreen pointer representing that motion. Ch 3

MP3 blog A blog where people post audio or music files. Ch 7

multicore processor A CPU chip that contains more than one processing unit (core); for example dual core (two cores) or quad core (four cores). Ch 3

multihomed device A device capable of connecting to a network in multiple ways. For example, your smartphone may be able to connect using either cell service or Wi-Fi. Ch 2

multimedia software Software that enables you to work with media, such as animation, audio, or video. Ch 5

multiprocessing When the OS allocates the work to be done amongst the cores or processors on a PC with two or more cores on a single CPU chip, or two or more CPUs. Ch 4

Multipurpose Internet Mail Extensions (MIME) format A format for messages that are sent over the internet. MIME permits text, graphics, audio, and video. Ch 2

multitasking The ability to have two or more tasks running at the same time. Also refers to the CPU's ability to execute several processes simultaneously. Ch 4

N

Near Field Communication (NFC) Technology that uses an RFID frequency for close-range communications to allow contactless payments, as with ApplePay. Ch 6

netbook A style of laptop computer with screen sizes ranging from 8 to 10 inches, and weighing only 2 to 3 pounds. Popular in the early 2000s, it was a precursor to tablets used for browsing the internet or using email. Ch 1

network adapter A device that provides the ability for a computer to connect to a network. Ch 6

network architecture The design and layout of the communications system; how computers in a network share resources. Ch 6

network attached storage (NAS) A networked hard drive. Ch 3

network interface card (NIC) One kind of network adapter card. In most current computers, NICs take the form of a circuit board built into the motherboard of a computer that enables a client computer on a LAN to connect to a network by managing the transmission of data and instructions received from the server. Ch 6

network operating system (NOS) Programs that control the flow of data among clients, restrict access to resources, and manage individual user accounts. Ch 6

network protocol A rule for how data is handled as it travels along a communications channel. Ch 6

node A device connected to a network. Ch 6

nonvolatile memory A type of computer storage specifically designed to retain information, even when the power is switched off. Ch 3

O

object linking and embedding (OLE) A technology that allows content to be treated as objects that can be inserted into different software documents, even if they were not created using that software. Ch 5

open content A creative work or other content that anybody can copy or edit online. Ch 7

open source Operating system software built with contributions by users whose source code is free to anybody to modify and use. Ch 4

open source software A type of software whose source code can be used, edited, and distributed by anybody. Ch 5

operating system (OS) A type of software that provides an interface for the user to interact with computer devices and software applications. Ch 4

operating system package Packaged software, such as Windows or Linux, which includes an operating system and utilities (collectively known as *system software*). Ch 4

optical drive A drive, such as a DVD or CD drive, that allows your computer to read and write data using optical technology. Ch 3

optoelectronic sensor Technology used in devices, such as optical drives, which detects changes in light caused by irregularities on a surface. Ch 3

organic light emitting diode (OLED) A display technology that projects light through an electroluminescent (a blue/red/green–emitting), thin film layer made of up of organic materials. Ch 3

output The information that results from computer processing or the act of writing or displaying such data. Ch 1

output device A device that allows a computer user to obtain data from a computing device; translates from electronic (digital) form to some other format. Ch 3

P

packet A small unit of data that is passed along a packet-switched network, such as the internet. Ch 6

packet switching The process of breaking data into packets, sending, and then reassembling the original data. Ch 6

parallelized Describes software design that allows pieces of a task to run on two or more processors at once. Ch 3

parallel processing A type of computer operation where several calculations or program instructions are carried out simultaneously. Ch 4

path The hierarchy of folders that leads to a stored file. Ch 4

PC card An add-on card that slots into a built-in card reader to provide other kinds of functionality, such as memory or networking capabilities. Ch 3

peer-to-peer (P2P) network A network architecture in which each computer in the network can act as both server and client. Ch 6

performance The speed with which your computer functions. Ch 4

peripheral device A device that physically or wirelessly connects to, and is controlled by, a computer but is external to the computer. Ch 3

petabit per second (Pbps) A transmission at a rate of 1 quadrillion bits per second. Ch 6

petaflop A measurement of supercomputing power representing a thousand trillion floating-point operations per second. Ch 1

phishing The practice of sending email that appears to be from a legitimate organization in an attempt to convince the reader to reveal personal information. Ch 8

photo editing software Software designed to enhance photo quality; often includes or special effects such as blurring elements or feathering the edges of a photo. Ch 5

photo printer An output device that allows you to print high-quality photos directly from a camera's flash memory to the printer without having to upload the photos to a computer first. Ch 3

photoblog A blog used to share amateur or professional photography. Ch 7

physical port A type of port that uses a physical cable to connect a computer to another device. Ch 3

piconet Networks that send or receive data among up to eight Bluetooth-connected devices. Also known as *personal area network (PAN)*. Ch 6

PIN (personal identification number) A set of numbers from 4 to 6 digits used as an alternative to a password to sign into an operating system, such as Microsoft Windows. Ch 8

pixel A single point in an image; short for picture element. Ch 1

platform The hardware architecture of a computer and the operating system that runs on it. Ch 4

platform dependency Applications and hardware that are only designed to work with a particular operating system. Ch 4

player A typically freely downloadable program that enables you to view or hear various types of online multimedia content. Ch 2

plotter An output device used to print large blueprints and other design or engineering documents. Ch 3

Plug and Play A feature that recognizes and makes available for use devices you plug into your computer, for example into a USB port. The OS installs the correct driver in order for the device to operate if the driver is available. Ch 4

plug-in A freely downloadable program that adds functionality to your browser. Ch 2

podcast A short audio presentation that can be posted online. Ch 5

port A slot in your computer used to connect it to other devices or a network. Ch 3

Post Office Protocol, Version 3 (POP3) A communications protocol that receives email from a mail server and delivers it to the proper mailbox. By default, messages are deleted from the server when the recipient retrieves his/her mail unless the user changes the settings. Ch 2

power supply Switches alternating current (AC) provided from a wall outlet to lower voltages in the form of direct current (DC). Ch 3

preferences Settings on your social networking page, including privacy settings. Ch 7

presentation software Software that enables you to create slideshows that include text, graphics, and multimedia. Ch 5

primary key field An Access database field that is used to ensure each row of data in a table is unique. Ch 5

printer A peripheral device used to produce printed output, or hard copies. Ch 3

private key A code key used in encryption that is known to only one or both parties when exchanging secure communications. Ch 8

processing The manipulation of data by the computer to create information. Ch 1

processor speed The speed at which the CPU interprets and carries out instructions that operate the computer and its devices. Ch 3

productivity software Software applications that people typically use to get work done; word processing, spreadsheet, database, or presentation software. This type of software is often compiled into suites of applications. Ch 5

profile A blogger or social networking user's information, such as name, location, and interests. Ch 7

public key A code key used in encryption. Creates an encrypted message that is decrypted by a private key. Ch 8

public key encryption A system of encrypting and decrypting data using a public key and private key combination. Ch 8

Q

query A question that can be used by database software to return information. Ch 5

quick response (QR) code A 2-D bar code that provides a shortcut to a website. Rather than entering the web address, you use your smartphone (with a reader application installed) to scan the code and let your phone's browser use the code to connect to the site. Ch 2

R

Radio Frequency Identification (RFID) A wireless technology primarily used to track and identify items using radio signals. An RFID tag placed in an item contains a transponder which is read by a transceiver or RFID reader. Ch 6

random access memory (RAM) A holding area for data while your computer processes information. When you turn your computer off, the data temporarily stored in this holding area disappears; RAM is therefore also referred to as *volatile memory*. Ch 1

ransomware A scam in which a user's computer is locked or data encrypted with a message from the malware creator demanding payment to restore access or data. Ch 8

read-only memory (ROM) Memory that holds information such as the BIOS and instructions the computer uses to start up the operating system. Also called *nonvolatile memory*. Ch 3

record A row of data in a table. Ch 5

relational database A database in which data is arranged into tables that are related on a common field. Ch 5

release to manufacturing (RTM) version A final version of the software program with all identified bugs reconciled so that the software can be duplicated and sold to the public or deployed to internal users. Ch 5

render farm A custom-designed connection between groups of computers joined in a computer cluster. Ch 1

repeater An electronic device that takes a signal and retransmits it at a higher power level to boost the transmission strength. A repeater can also transmit a signal to move past an obstruction, so that the signal can be sent further without degradation. Ch 6

resolution A measurement of the number of pixels on a screen. Ch 1

responsive design Design that allows web pages to sense the device being used and reformat the content for optimum display on all screen sizes. Ch 5

RFID reader An input device often used in retail or manufacturing settings to scan an embedded tag using radio frequency. Ch 3

rich text format (RTF) A text format that includes only basic formatting information that most software products are likely to be able to open or import. Ch 5

ring topology An arrangement that has computers and other devices one after the other in a closed loop. Data transmitted on a ring network travels from one computer to the other until it reaches its destination. Ch 6

rootkit A set of programs or utilities designed to allow a hacker to control a victim computer's hardware and software and permit a hacker to monitor the user's actions. Ch 8

router A hardware device that connects two or more networks in a business setting. At home, a router allows you to connect multiple devices to one high-speed connection. Ch 6

S

satellite communication Space-based equipment that receives microwave signals from an earth-based station and then broadcasts the signals back to another Earth-based station. Ch 6

scanner A peripheral input device used to create an electronic file from a hard copy document. Ch 3

scareware A scam where an online warning or pop-up convinces a user that his or her computer or mobile device is infected with malware or has another problem that can be fixed by purchasing and downloading software, which may do nothing or install malware. Ch 8

screen capture software Software that enables you to capture an entire computer screen or only a portion of it. Ch 5

SDRAM (synchronous dynamic random access memory) An updated version of DRAM that provides significant improvements in access speed. Most modern computer memory is some variation of SDRAM, including DDR-SDRAM, DDR2-SDRAM, and DDR3-SDRAM. Ch 1

search directory A site that allows you to locate web content within categories. Ch 2

search engine A website that permits you to search for information by entering keywords. Ch 2

Secure Socket Layer (SSL) A cryptographic protocol that is required for creating Transport Layer Security (TLS). Ch 8

Semantic Web The next (third) phase in online usage, also called *Web 3.0*, which will make it possible for websites to "understand" the relationships between elements of web content. Ch 2

serial port A port built into the computer that is used to connect a peripheral device to the serial bus, typically by means of a plug with 9 pins. Ch 3

server Any combination of hardware and software that provides a service, such as storing data, to a client, such as your computer. Ch 6

shared feature A small application that cannot run on its own, but that allows suites of software products to share functionality, such as diagramming or drawing. Ch 5

shareware Software for which you pay a small fee. Ch 5

Short Message Service (SMS) A service that carries text messages and is used by cell phone providers. Ch 2

Simple Mail Transfer Protocol (SMTP) A communications protocol installed on the ISP's or online service's mail server that determines how each message is to be routed through the internet and then sends the message. Ch 2

sketchblog A blog consisting of drawings and sketches. Ch 7

smartphone Mobile phone devices with a mobile OS and rich feature set that essentially makes them into very small computers. Ch 4

smart speaker A wireless, voice-controlled device with an integrated virtual assistant (artificial intelligence) that offers interactive actions. Ch 2

social bookmarking A method of sharing bookmarks with others using tags. Ch 7

social engineer A con artist who employs tactics to trick computer users into giving up valuable information. Ch 8

social journaling A form of blogging where brief comments rather than personal blogs are the main form of interaction, as on Twitter. Also called *microblogging*. Ch 7

social mobile media Social services accessed from mobile phones. Ch 7

social networking site A site that offers the ability to share contacts and build a network of "friends," along with tools that allow individuals and groups to connect and communicate. Ch 7

social web The collective description of websites that offer the ability to communicate, interact, and network with others. Ch 7

software as a service (SaaS) A software delivery model where a provider licenses an application to customers to use as a service on demand. Ch 5

software development life cycle (SDLC) The general flow of creating a new software product; includes performing market research and business analysis, creating a plan and budget for implementing the software, programming the software, testing the software, releasing the software to the public, and debugging the software. Ch 5

software engineering (SE) A field involving writing software programs, which might be developed for a software manufacturer to sell to the public, or involve a custom program written for a large organization to use in-house. Ch 1

software on demand A software delivery model where a provider licenses an application to customers to use as a service on demand. Also called *software as a service (SaaS)*. Ch 5

software suite A collection of productivity software applications sold as one package that use tools common to all the products in the suite. Ch 5

solid-state drive (SSD) A flash-based replacement for an internal hard disk that is lighter and more durable than a traditional hard disk; should pave the way for smaller, more portable computers with longer battery life. Ch 3

source code The programming code used to build a software product. Ch 5

spam Mass emails sent to those who haven't requested them, usually for the purpose of advertising or fraud. Ch 8

speaker A device that provides audio output. Ch 3

spear-phishing A targeted phishing attack sent to individuals employed by certain companies for the purpose of obtaining trade secrets or other confidential information, leading to financial gain for the hacker. Ch 8

speech recognition software A type of software that enables a computer to recognize human speech and convert it into text. Ch 5

spider An intelligent agent that creates entries for a search engine index based on words found on web pages. Ch 2

spoofing Attempting to gain valuable information via electronic communications by misleading a user as to your identity. Ch 8

spreadsheet software An application with which you can perform calculations on numbers and display other data. In addition, most spreadsheet products offer sophisticated charting and graphing capabilities. Ch 5

spyware Software that tracks activities of a computer user without the user's knowledge. Ch 8

SRAM (static random access memory) A type of memory that is about five times faster than DRAM. Though dependent upon electricity, this type of memory does not require constant refreshing and is more expensive than DRAM. It is therefore often used only in cache memory applications. Ch 1

standards Allow different devices to talk to one another. Standards ensure compatibility among devices, specifying how computers access transmission media, the speeds used on networks, the design of networking hardware such as cables, and so on. Ch 6

star topology An arrangement where all the devices on the network, called *nodes*, connect to a central device that is a hub or a switch. Ch 6

Start menu A menu that appears when you click the Start button on the Windows taskbar that displays commands, apps menus, and rectangular tiles used to launch apps, the web, documents, or programs. Ch 4

storage A permanent recording of information, data, and programs on a computer's storage medium, such as a magnetic disk or optical disc, so that they can be retrieved as needed. Ch 1

streaming media Media that is delivered to your computer as a constant stream of content, usually requiring a media player. Ch 2

strong password A password that is difficult to break. Strong passwords should contain uppercase and lowercase letters, numbers, and punctuation symbols, but not contain dictionary words or repeating characters. Ch 8

Structured Query Language (SQL) The standard language used to query a database. Pronounced "sea-quel." Ch 5

stylus A special device, usually with a rubber tip, that is used to tap on a touchscreen. Ch 3

supercomputer A computer with the ability to perform trillions of calculations per second, usually custom-made for a particular use or used as a large server. Ch 1

surface-conduction electron-emitter display (SED) A display technology that uses nanoscopic electron emitters (extremely tiny wires smaller than human hairs) to send electrons that illuminate a thin screen. Ch 3

surge protector Protects an individual device from loss of data caused by a spike in power, such as might occur during a thunderstorm. Ch 8

swap file A file created when data is stored or "swapped" into virtual memory. Ch 4

switch A hardware device that joins several computers together to coordinate message traffic in one LAN network. Although a switch performs a role similar to a hub, a switch checks the data in a packet it receives and sends the packet to the correct destination using the fastest route. Ch 6

symmetric encryption A system of encrypting and decrypting data where in the sending and receiving computers each have a matching private key. Ch 8

sync The process of updating data on one device based on changes made to the data on another device. Short for synchronize. Ch 2

system configuration The entire computing system, including the identity of the computer, the devices connected to it, and some essential processes that the computer runs. Ch 4

system files Files that provide instructions needed to run the operating system on your computer. Ch 4

system software Software that includes the operating system and utilities for maintaining a computer and its performance. Ch 4

T

table A row-by-column layout of data. Ch 5

tablet A portable computer that enables you to give commands via a touchscreen using easy controls. Tablets also enable you to add functionality via downloadable apps and are often used as e-readers and media-consumption devices. Ch 1

tags Labels assigned to blog posts that describe the various topics covered in each post so that users can easily search a blog by topic. Ch 7

TCP/IP Short for Transmission Control Protocol/Internet Protocol. A protocol that breaks transmissions into small packets of data that are sent on the network. Each packet specifies the order in which the data is to be reassembled. This is the protocol for the internet. Ch 6

technological convergence When a device begins to use technologies traditionally thought to belong to another device, as when a cell phone performs tasks traditionally performed by a computer. Ch 1

terabit per second (Tbps) A transmission at a rate of 1 trillion bits per second. Ch 6

tethering The ability to share the internet connection of a mobile device with another device via a cable, Bluetooth, or Wi-Fi. Ch 2

text message A brief, written message of 160 characters or less sent between mobile phones or other portable devices using the short message service (SMS). Ch 2

texting The process of sending a text message. Ch 2

thermal printer A type of printer that heats coated paper to produce output. Ch 3

thin film transistor active matrix liquid crystal display (TFT active matrix LCD) The most prevalent type of monitor technology used today; uses a thin film transistor (TFT) to display your computer's contents. Ch 3

Thunderbolt port A port connection technology introduced by Apple that provides a high-speed serial interface using multiple channels to transmit data; used on devices such as cameras, camcorders, and audio and video equipment. Ch 3

top-level domain (TLD) The suffix (the period and the letters that follow) of a domain name. Ch 2

topology How devices in a network are physically arranged and connected to each other. Ch 6

touchpad A type of flat mouse or pointing device that senses finger movement; often used in laptop computers. Ch 3

touchscreen A visual display that permits the user to interact with a digital device by touching various areas of the screen to provide input and viewing information onscreen as output. Ch 3

Transport Layer Security (TLS) A protocol that protects data, such as credit card numbers as they are being transmitted between a customer and online vendor or payment company. Ch 8

tree topology A hybrid of the star and bus topology, tree topology involves two or more star networks joined via a main bus cable. Ch 6

Trojan horse Malware that masquerades as a useful program. When you run the seemingly useful program, you let this type of malware into your system. It opens a "back door" through which hackers can access your computer. Ch 8

twisted-pair cable A type of cable consisting of two independently insulated wires twisted around one another. This type of cable is used to transmit signals over short distances. Twisted-pair cables are used to connect a home's hardware telephone system or an Ethernet network, for example. Ch 6

two-factor authentication (2FA) A form of authentication that adds a second layer of security. The most commonly used 2FA is typing your password and then receiving a code via text message that you enter on a second screen. Also referred to as *two-step verification*. Ch 8

U

ubiquitous computing Placing computing power in your environment as with, for example, a system in your house that senses and adjusts lighting or temperature. Also called *embedded technology*. Ch 1

UEFI (Unified Extensible Firmware Interface) A new specification for booting your computer that will eventually replace the aging BIOS firmware and could make booting computers a much faster process. Ch 4

ultrabook A type of lightweight laptop. Ch 1

Unicode An encoding standard used to represent different languages and scripts by assigning each letter, digit, or symbol a unique numeric value. This value is applied across different platforms and programs and is recognized internationally. Ch 1

uniform resource locator (URL) A naming system used to designate unique website addresses that you enter into a browser to navigate to a particular site. Also called *web address*. Ch 2

uninterruptible power supply (UPS) A battery backup that provides a temporary power supply in case of a power failure. Ch 8

universal serial bus (USB) port A port in the form of a small rectangular slot that can be used to attach everything from wireless mouse and keyboard toggles (the small device that transmits a wireless signal to a wireless device) to USB flash drives for storing data. Ch 3

UNIX A server operating system written with the C programming language. Ch 4

upload To transmit data, such as a digitized text file, sound, or picture from one's own computer to a remote site via a network, such as the internet. If you are sending content to the internet, you are uploading. Ch 2

user interface The visual appearance that software presents to a user. Ch 4

utility software A category of system software that you use to optimize and maintain your system performance and provides information about system resources. Ch 4

V

video blog A blog where people post video content. Also called *vlog*. Ch 7

video conferencing Technology that transfers video and audio signals over the internet so users can see as well as hear one another. Ch 2

video editing software Software used to create and edit video files. Ch 5

virtual memory A capability of the computer's operating system that handles data that cannot fit into RAM when running several programs at once. When RAM is used up, data is stored or "swapped" into virtual memory. Ch 4

virtual private network (VPN) A connection to a corporate network via a VPN Server. Employees who need to access the company's private network from a public location, such as an airport or hotel, connect using a VPN. Ch 6

virtual reality system A system that connects you to a simulated world. It creates a connection between user and computer that allows both input and output in various forms and can be used to create sophisticated training programs such as those used by pilots, doctors, and astronauts. Ch 3

virus A type of computer program that is placed on your computer without your knowledge. The key characteristic of a virus is that it can reproduce itself and spread from computer to computer by attaching itself to another, seemingly innocent, file. Ch 8

virus definitions Information about viruses used to update antivirus software to recognize the latest threats. Ch 8

Voice over Internet Protocol (VoIP) A transmission technology that allows you to make voice calls over the internet using a service such as Skype. Ch 2

volatile memory A type of computer memory whereby stored instructions and data are lost if the power is switched off. Ch 1

W

warm boot Restarting a computer without turning the power off. Ch 4

warm server A server activated periodically to get backup files from the main server. Ch 8

wearable computer An electronic device that is worn and provides computing functionality in the form of everything from clothing to glasses or watches. Ch 1

web (World Wide Web or WWW) A system that contains the body of content available as web pages that are stored on internet servers. Ch 2

Web 2.0 An evolution of the web, considered its second phase, associated with users not only reading content, but interacting by writing content. Provides users with a means to collaborate—share, exchange ideas, and add or edit content. Ch 2

Web 3.0 The next (third) phase in online usage, also called the *Semantic Web*, which will make it possible for websites to "understand" the relationships between elements of web content. Ch 2

web address A naming system used to designate unique website addresses that you enter into a browser to navigate to a particular site. Also called *uniform resource locator (URL)*. Ch 2

web authoring software A type of software that provides tools for creating and editing web pages. Ch 5

web-based email A process that allows you to create and send email messages, add attachments to email messages, store contact information, and manage email messages in folders. Ch 2

web-based software Software that is hosted on an online provider's website; you access it over the internet using your browser. This type of software is not installed on your computer. Ch 5

web-based training Learning that typically contains a self-directed element and takes place via the web using some combination of text, multimedia, and interactive tools. Ch 5

webcam Video cameras that are either built into your computer monitor or purchased separately and mounted on to a computer. Ch 3

web conferencing A combination of technologies that allow you to hold meetings online with voice, video, images, and various collaborative tools such as virtual whiteboards. Ch 2

web development A technology career that includes diverse web-based activities, including programming and developing websites, developing text and visual content for websites, and using social media to promote goods and services. Ch 1

web page A single document on the web, which may contain text, images, interactive animations, games, music, and more. Several related web pages make up a website. Ch 2

website A collection of related web pages. Ch 2

wide area network (WAN) A type of network that serves larger geographic areas. WANs, such as the internet, are used to share data between networks around the world. WANs might use leased T1/T3 lines, satellite connections, radio waves, or a combination of communications media. Ch 6

Wi-Fi A wireless technology, based on the 802.11 standard in its various versions such as 802.11 a, g, or n, that is used to connect to the internet via hotspots and radio waves. Ch 6

Wi-Fi Protected Access (WPA) An encryption standard used to protect data sent across a wireless network. Designed by the Wi-Fi Alliance to overcome the security limitations of Wired Equivalent Privacy (WEP). Ch 8

Wi-Fi Protected Access 2 (WPA2) A stronger, more complex encryption standard than WPA. WPA2 became mandatory in 2006 for all new equipment certified by the Wi-Fi Alliance. Ch 8

Wi-Fi Protected Access 3 (WPA3) WPA3, which increases the security of home and public networks, is the latest encryption standard announced by the Wi-Fi Alliance in 2018. Ch 8

wiki Online content libraries, such encyclopedias or dictionaries, that allow any user to contribute and edit content. Ch 7

Windows 10 The latest Microsoft Windows OS, released beginning in 2015, which introduced an intelligent assistant search tool named Cortana and the Microsoft Edge browser. Users can interact with their desktop, laptop, or tablet using a touchscreen. The Start menu contains live tiles that represent apps and can update on the fly. Ch 4

wired data gloves Equipment worn on the hands that allows users to communicate with a virtual reality system. Ch 3

Wired Equivalent Privacy (WEP) An encryption standard used to protect data sent across a wireless network. An older and less secure technology than Wi-Fi Protected Access (WPA). Ch 8

wireless access point A hardware device that contains a high-quality antenna that allows computers and mobile devices to transmit data to each other or to and from a wired network infrastructure. Ch 6

wireless adapter A piece of equipment used to connect a computer to wireless networks. Ch 3

Wireless Application Protocol (WAP) Specifies how mobile devices such as cell phones display online information, such as maps and email. Ch 6

wireless charging A charging method that uses a charging station to transfer an electromagnetic charge to a device. Ch 3

wireless interface card A network interface card that uses wireless technology. Ch 6

wireless LAN (WLAN) A local area network that uses wireless technology. Ch 6

wireless modem A piece of hardware that typically takes the form of a PC card that you slot into a device to provide it with an antenna that can pick up a connection to the internet. Ch 6

wireless router A hardware device that allows you to connect multiple networks using wireless communication signals. Ch 6

word processor software A type of software used to create documents that include sophisticated text formatting, tables, photos, drawings, and links to online content. Ch 5

worm A self-replicating computer program with the ability to send out copies of itself to every computer on a network. Worms are usually designed to damage the network, in many cases by simply clogging up the network's bandwidth and slowing its performance. Ch 8

WYSIWYG (what you see is what you get) Pronounced "wiz-e-wig." A feature of web authoring software that enables it to show onscreen exactly how the file will look when displayed in a web browser. Ch 5

Z

zombie A computer compromised by malware that becomes part of a botnet and is used to damage or compromise other computers. Also called a *bot*. Ch 8

Index

time spent on, 194
uses of, 26
two-factor authentication (2FA), 211

U

Uber, security breach, 208
ubiquitous computing, 7
Ubuntu, 76, 95, 166
UEFI (Unified Extensible Firmware Interface), 86–87, 88
ultrabooks, 4
Unicode, 11–12
uniform resource locator (URL), 37
uninterruptible power supply (UPS), 231
UNIVAC I, 89
universal serial bus (USB), 63
UNIX, 89–90, 91, 94
upload, 36
USB (flash) drives, 15, 66
Usenet, 190
user authentication, 228–229
user interfaces, 97
utility software, 86, 105

V

video blogs, 189
video conferencing, 50
video editing software, 126–127
virtual memory, 100, 101
virtual private network (VPN), 47, 168
virtual reality systems, 73
virus definitions, 232
viruses, 213–214, 232
VisiCalc, 118
vlogs, 189
Voice over Internet Protocol (VoIP), 50
VoiceThread, 199
volatile memory, 12

W

WannaCry ransomware, 208
WANs (wide area networks), 160
warm boot, 86
warm server, 230
watchOS, 108
WBAN (wireless body area network), 69
wearable computers, 3
    data privacy and, 226
wearable devices, 8, 28, 69, 108
Wear OS, 108
web. *See also* internet
    deep, 44–45
    described, 29
    evaluating content, 42–43
    invisible, 44–45
    navigating, 37–39
    phases of, 31–33
    searching, 39–42, 44–45
    three Ws of, 43
Web 2.0, 31–32, 169. *See also* social web
Web 3.0, 31, 32–33
web address, 37–38
web authoring software, 127

web-based email, 48
web-based software, 134. *See also* cloud-based software
web-based training, 127–128
webcams, 68–69
web conferencing, 50
web development, 9
web page, described, 29, 30
websites. *See also* specific sites
    browsing, 38–39
    cookies, 218–219
    described, 29, 30
    IPs and URLs, 37–38
    parts of web page, 38–39
    recognizing secure, 218
    searching for content on, 39–42
    as specialize search sites, 41–42
    three Ws of, 43
WEP (Wired Equivalent Privacy), 213
WeVideo.com, 198
wide area networks (WANs), 160
Wi-Fi
    connecting to internet, 35
    man-in-the-middle (MITM) attack, 222–223
    networks, 35, 157
    using safely, 222–223
Wi-Fi Protected Access (WPA), 213
Wi-Fi Protected Access 2 (WPA2), 213
Wi-Fi Protected Access 3 (WPA3), 213
Wii U, 69
Wikipedia, 32, 43, 182, 183, 197
wikis, 196–197
Wikitravel, 197
Wiktionary, 197
Windows
    with ARM, 4
    botnets and, 216
    Defender, 127
    Disk Cleanup, 86
    drives, 65
    File Explorer, 103
    file management, 103
    Help features, 104
    history of, 93
    malware and, 212
    operating system, 3, 76, 90, 91, 92, 93
    PIN for, 234
    10, Windows, 93, 97, 127
    updates for, 232–233
wired data gloves, 69
Wired Equivalent Privacy (WEP), 213
wired networks
    standards for, 156
    transmissions systems for, 152–153
wired transmissions, 152–153, 156
wireless access points, 34, 165–166
wireless adapters, 65
Wireless Application Protocol (WAP), 159
wireless body area network (WBAN), 69
wireless charging, 63
wireless interface cards, 164
wireless LANs (WLANs), 160
wireless mobile storage devices, 66

wireless modems, 35, 164
wireless networks
    standards for, 156–159
    transmissions systems for, 153–155
wireless personal area network (WPAN), 69
wireless routers, 165
wireless transmissions, 153–155, 156–159
WLAN (wireless LAN), 160
WolframAlpha search engine, 33
Word, 120
    Office 365, 169
WordArt, 138
WordPad, 88
WordPress, 135
word processor software, 120
WordStar, 118
work, computer security at, 227–231
World Intellectual Property Organization
        (WIPO), 44

World Wide Web Consortium (W3C), 127
worms, 214
WPAN (wireless personal area network), 69
WYSIWYG (what you see is what you get), 127

X

Xbox One, 69
Xerox, 71

Y

YouTube, 190, 193
    as media sharing site, 198
    time spent on, 194

Z

Zappos, 45
ZDNet, 187
zombies, 216

Photo Credits

Microsoft is a trademark or registered trademark of Microsoft Corporation in the United States and/or other countries. Microsoft images used with permission from Microsoft.

Chapter 1
Page 2, courtesy of Apple, Inc., © Shutterstock.com/makalex69, © Shutterstock.com/robtek, © Shutterstock.com/Hadrian; *page 4*, courtesy of Apple, Inc.; *page 5*, © iStock.com/Saturated; *page 7*, © iStock.com/metamorworks; *page 9*, © iStock.com/baranozdemir; *page 10*, © iStock.com/bunhill; *page 11*, courtesy of Xerox Corporation; *page 12*, © Shutterstock.com/Daniel Chetroni; *page 13*, (top) Adobe product screenshot reprinted with permission from Adobe Systems Incorporated. Computer image courtesy of Dell, Inc.; *page 14*, (left) © Shutterstock.com/kavring, (right) © Shutterstock.com/cigdem.

Chapter 2
Page 25, © Shutterstock.com/zffoto; *Page 26*, © iStock.com/STEEX; *page 27*, (top) © Shutterstock.com/VDB Photos, (bottom); *page 28*, © Simon Jarratt/Corbis; *page 35*, © Shutterstock.com/zeljkodan; *page 36*, (top) used with permission from Opera; *page 43*, (bottom) The Wikipedia unified mark is a trademark of the Wikimedia Foundation and is used with the permission of the Wikimedia Foundation. We are not endorsed by or affiliated with the Wikimedia Foundation.; *page 46*, (bottom) © 2014 EBAY INC. ALL RIGHTS RESERVED, reproduced with the permission of eBay Inc.; *page 48*, © Daniel L. Murphy; *page 50*, © Shutterstock.com/Andrey_Popov.

Chapter 3
Page 60, (clockwise from top left) © iStock.com/Saturated, courtesy of Sony Corporation of America, © Shutterstock.com/Hadrian, courtesy of Hewlett-Packard Company, © iStock.com/spooh, © Shutterstock.com/pianodiaphragm; *page 61*, courtesy of ASUSTeK Computer Inc.; *page 63*, © Shutterstock.com/Thanaphat Kingkaew; *page 66*, © iStock.com/calvio; *page 67*, (top) courtesy of V-MODA, (bottom) courtesy of Hewlett-Packard Company; *page 68*, (top to bottom) ©Shutterstock.com/

blackzheep, © iStock.com/Wavebreakmedia, © iStock.com/alexxl66; *page 69* (top) © iStock.com/RamonCarretero, (bottom) courtesy of National Science Foundation; *page 71*, (clockwise from top left) © iStock.com/AJ_Watt, © iStock.com/adventtr, courtesy of Panasonic; *page 72*, Shutterstock.com/Brian A Jackson; *page 73*, (top) courtesy SMART Technologies, (bottom) © iStock.com/Imgorthand; *page 75*, © Shutterstock.com/Adam Ziaja.

Chapter 4
Page 87, courtesy of Apple, Inc., *page 89*, (bottom) courtesy of Hagley Museum and Library; *page 90*, courtesy of Apple Computers; *page 92*, (top to bottom) courtesy of Microsoft, courtesy of Apple, Creative Commons Wikipedia/The fb209; *page 94*, courtesy of Apple, Inc., *page 95*, courtesy of Linux Online, Inc; *page 96*, (top and bottom) courtesy of Google; *page 98*, (bottom) courtesy of Hewlett-Packard Company; *page 100*, (bottom) courtesy of Apple, Inc., *page 104*, (top) courtesy of Apple, Inc., *page 107*, (left to right) © Shutterstock.com/Framesira, courtesy of LG, © Shutterstock.com/DR-images.

Chapter 5
Page 124, courtesy of Apple, Inc., *page 126*, (clockwise, starting in top left) Adobe Creative Cloud reprinted with permission from Adobe Systems Incorporated, courtesy of Quark Software Inc., used with permission from Corel Draw, TechSmith Snagit logo reprinted with permission from TechSmith Incorporated *page 127*, courtesy of Adobe Systems Incorporated; *page 128*, courtesy of gamesforchange.org; *page 129*, all courtesy Intuit Inc; *page 130*, © iStock.com/Adivin.

Chapter 6
Page 150, © Shutterstock.com/Alexlukin; *page 155*, © iStock.com/cpku; *page 158*, © Shutterstock.com/metamorworks; *page 164*, (top) © iStock.com/Alec051, (bottom) © iStock.com/baloon111; *page 165*, (clockwise from top left) courtesy of Cisco, courtesy of TRENDnet, courtesy of Amped Wireless, courtesy of TRENDnet; *page 171*, © Shutterstock.com/Zapp2Photo.

Chapter 7

Page 183, (top) © iStock.com/jabejon, (middle) © iStock.com/bo1982, (bottom) © Shutterstock.com/wavebreakmedia; *page 185*, © iStock.com/atakan; *page 186*, courtesy of Khan Academy; *page 187*, courtesy of HootSuite Media Inc.; *page 188*, courtesy of DOI.gov; *page 190*, courtesy of Tumblr, Inc.; *page 191*, courtesy of Facebook; *page 192*, (top) © Shutterstock.com/PixieMe, (bottom) © Shutterstock.com/Sean Locke Photography; *page 195*, courtesy of Flipora; *page 196*, courtesy of www.pearltrees.com; *page 197*, courtesy of wikia.com; *page 198*, © Shutterstock/Sirirat; *page 199*, courtesy of SoundCloud.

Chapter 8

Page 208, © Shutterstock.com/Jarretera; *page 211*, courtesy of McAfee; *page 212*, photo courtesy of Linksys; *page 217*, image used with permission from MalwareBytes; *page 218*, (clockwise from upper right) courtesy of TRUSTe Inc, courtesy of Validatedsite.com, Copyright © 2014 Symantec Corporation. All rights reserved. Reprinted with permission from Symantec Corporation; *page 219*, (top right) courtesy of Apple, Inc., (bottom) © iStock.com/mbolina; *page 222*, (upper left) courtesy of Kensington Computer Products Group, a division of ACCO brands, (upper right) © Shutterstock.com/ptnphoto, (bottom) © Shutterstock.com/Kzenon; *page 223*, © iStock.com/sturti; *page 224*, © iStock.com/metamorworks; *page 225*, © Shutterstock.com/Vladmir Badaev; *page 226*, © Shutterstock.com/Tyler Olson; *page 228*, © Shutterstock.com/Titkul_B; *page 229*, © Shutterstock.com/Tanoy1412; *page 230*, © iStock.com/JoeBiafore; *page 233*, © AVG Netherlands B.V. and the AVG group of companies.